T0260065

EUROPE'S DRAGONFLIES

A field guide to the damselflies and dragonflies

Dave Smallshire and Andy Swash

WILDGuides

PRINCETON
press.princeton.edu

Published by Princeton University Press,
41 William Street, Princeton, New Jersey 08540
In the United Kingdom: Princeton University Press, 6 Oxford Street,
Woodstock, Oxfordshire OX20 1TR
press.princeton.edu

First published 2020

British Library Cataloging-in-Publication Data is available

Library of Congress Control Number 2020931167
ISBN 978-0-691-16895-1

Production and design by **WILD**Guides Ltd., Old Basing, Hampshire UK.
Printed in Italy

10 9 8 7 6 5 4 3 2 1

Contents

Introduction 4

How to use this book 6

The types of dragonfly 8

    Damselflies (Zygoptera) 8

    Dragonfies (Anisoptera) 10

How to identify dragonflies 12

Glossary 14

THE SPECIES ACCOUNTS 15

Introduction to damselflies (Zygoptera) 16

    Spreadwings & winter damsels 18

    Demoiselles 38

    Odalisque 48

    Red damsels 50

    Featherlegs 58

    Bluets 64

    'Blue-tailed' damselflies 94

Introduction to dragonflies (Anisoptera) 122

    Hawkers, spectres & emperors 126

    Clubtails, snaketails, pincertails, hooktail & Bladetail 176

    Goldenrings, cruiser & cascader 208

    Emeralds 230

    Baskettail & chasers 250

    Skimmers 258

    Whitefaces 282

    Darters, scarlet, dropwings, pennant, percher & groundling ... 294

    Gliders 338

Conservation status & legislation 342

Checklist of species 343

Further information 348

Acknowledgements & photographic credits 349

Index 356

Introduction

Dragonflies are among the most attractive and amazing of all insects. Their large size (20–96 mm long) makes them relatively easy to study and they have become increasingly popular in Europe. Along with butterflies, dragonflies are often colourful, and dazzling, and have considerable aesthetic appeal. They have excellent vision and powers of flight and unique breeding behaviour.

Dragonflies is the term commonly used, including in this book, for the insect order Odonata. This order comprises both damselflies and dragonflies (often referred to as odonates). With over 6,300 dragonfly species described, this makes the 140 species in Europe (as defined in this book) a modest total, and a manageable group of insects to study.

The body of a dragonfly has three parts: head, thorax and abdomen. The head is dominated by a pair of large compound eyes but also has small antennae and complex mouthparts adapted to keep prey secure while chewing. The short thorax has attached to it three pairs of jointed, spiny legs and two pairs of intricately veined wings. The long, typically slender, abdomen contains the reproductive organs, which in males comprise appendages at the tip and secondary genitalia on the underside near the front. Females lay eggs through an opening underneath the tip; in some species, this takes the form of a robust ovipositor that is used to inject eggs into plant material.

In general, Europe's dragonflies spend most of their lives as eggs and aquatic larvae. Although the relatively short-lived adults are mostly seen at water, they are by no means restricted to wetlands or waterside habitats. After emergence they continue their lives as carnivorous predators and may wander far from the freshwater or brackish sites that were their pre-adult home – sometimes making long migratory flights. As cold-blooded animals, the flight period of most species is during the warmer months of April to September, although some may be found flying at other times, especially in southern Europe.

This book focuses on the field identification of adult dragonflies that breed or have occurred naturally

COMMON DARTER
Sympetrum striolatum

Dragonflies have huge compound eyes that provide almost all-round vision in colour, including ultraviolet and polarized light, and superior motion-detection – all vital adaptations for detecting potential predators, mates, rivals and prey.

BLUE-EYED HAWKER
Aeshna affinis

Dragonfly wings are remarkably strong and light, but flexible, with a complex, highly evolved structure. They are powered by strong muscles in the thorax, with neurons connected directly to the brain. Bursts of speed can reach 15 m/s, while species with a large wing area can migrate up to 6,000 km using tail winds.

VIOLET DROPWING
Trithemis annulata

Colours are produced by light reflecting from particles within or on the cuticle. Some species are pruinose, as a result of a waxy secretion; others have shiny, 'metallic' bodies and/or wings.

in Europe (accidental introductions are not included). It is aimed at both those with expertise visiting a new area that may have unfamiliar species, as well as those with less experience who are seeking to increase their knowledge. It avoids the use of technical terms wherever possible for ease of interpretation.

BLADETAIL
Lindenia tetraphylla

Adults feed mainly on flying insects, which are collected in a 'basket' of spiny legs, each ending in a pair of sharp claws. Damselflies may also pluck small invertebrates from vegetation.

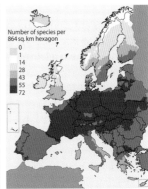

Number of species per 864 sq. km hexagon
0
1
14
28
43
55
72

This map shows the species richness of damselflies and dragonflies across Europe, based on a map shown in the *European Red List of Dragonflies* by Kalkman *et al.* (2010) (see *p. 342*).

LARGE RED DAMSEL
Pyrrhosoma nymphula

Emergence of adults from the larval skin typically lasts for a few hazardous hours. Colours change as the chitinous exoskeleton hardens, and again with sexual maturity and old age.

BANDED DARTER
Sympetrum pedemontanum

Uniquely, males use abdominal appendages to grasp females behind the head or on the pronotum, often linking in flight.

HAIRY HAWKER
Brachytron pratense

The female of a 'tandem' pair may link the tip of her abdomen to secondary genitalia under the base of the male's abdomen, forming a mating 'wheel'. Some species mate for a few seconds in flight; others take many minutes or even hours while perched.

BLUE FEATHERLEG
Platycnemis pennipes

Eggs are laid into plant material or organic or mineral deposits, or flicked into water or damp ground. Males may or may not stay in attendance, or may remain attached to the female.

How to use this book

This is an identification guide to adult dragonflies, based on individual accounts for each of the species (including all of the currently recognized subspecies and variations) recorded in Europe up to the end of 2018. The 140 species can be conveniently divided into 16 broad 'types' (seven of damselfly and nine of dragonfly) based on their general appearance. Images of each of these 'types' are shown at approximately life-size on the following four pages, cross-referenced to the relevant introductory sections, which provide more details of the key identification features and are in turn cross-referenced to the appropriate species account(s). This approach provides options to navigate through the book step-by-step via different routes, depending on your level of knowledge and expertise. Brief guidance on identifying dragonflies is provided, along with illustrations showing their key features. A glossary (*p. 14*) of the technical terms used precedes a key (*p. 15*) to the format of the species accounts that follow.

Species that look similar are grouped close to each other in the species accounts, rather than using a strict taxonomic sequence. The descriptions are as consistent as possible and use standardized terminology. They start with a summary of key features that are shared by both sexes and then give the features that are important for the identification of males and females, and finally any variations, including recognized colour and geographical forms and those that are related to age. These are presented in a consistent order and key features are highlighted in **bold** text; in combination, these features are diagnostic of the species. The species accounts also include information on behaviour and habitat preferences that may be helpful in identifying a species. For each species there is a photo showing typical habitat. Summary information is also included about legal protection, status, population trends, habitat type(s) and flight periods, sub-divided where appropriate to show differences between northern and southern Europe. The ranges in overall length and hindwing length are given. The maximum and minimum lengths are indicated graphically using a scale bar on the left of the page, shown at life-size. Scale bars are also included in the introductory sections for each 'type'. In the descriptions, size is described relative to all damselflies or all dragonflies. Finally, a box lists 'LOOK-ALIKES': species that might be confused with the species in question. Only those that overlap in range and habitat of that species are included.

Carefully selected photographic images are used to illustrate both sexes of each species, often together with additional images showing variations due to immaturity or old age, female colour forms and geographical variations. The images on each plate are scaled at either life-size ('large' dragonflies), 1·3× life-size (baskettail and chasers), 1·5× life-size (other dragonflies) or 2× life-size (all damselflies). The plates and species accounts are supplemented by images or graphic illustrations of critical identification features, including those that in a small number of cases can only be seen in the hand or possibly in high quality photographs. Mature dragonflies can be caught with a net and handled safely, but it should be remembered that rare species may have legal protection in certain countries and may therefore need permission to be caught and handled. Those species protected under the EU Habitats Directive and/or the Bern Convention (see *Conservation and legislation, p. 342*) are highlighted in a complete *Checklist of species* (*p. 343*), which includes all the species recorded naturally in Europe.

The vernacular (English) and scientific names follow the *Guide to the Dragonflies of Britain and Europe* (see *Further information, p. 348*); other English names commonly used in Europe are included in the main species accounts. When referring to a particular species these are shown with initial capitals (*e.g.* Blue Hawker); references to groups of species are shown in lower case text (*e.g.* hawkers).

Distribution maps

A map showing the distribution of records since 1990 is included for each species, based on an interpretation of the maps in the *Atlas of the European dragonflies and damselflies* (see *Further information, p. 348*). It should be noted that some regions are not particularly well surveyed and records are sparse; the maps may therefore be imprecise especially in parts of eastern Europe. It is important to remember that species with more demanding habitat requirements may occur very locally within an overall range, so the shading covers areas of unsuitable breeding habitat as well as the locations of migrant records; for these reasons the maps should be regarded as indicative. Recent significant range expansions have been included where possible. For the purpose of this book, Europe is defined as shown in the map below. This includes the popular tourist destinations of the Macaronesian and Mediterranean islands (where they belong to European countries) and western Turkey. The eastern boundary is pragmatic and omits areas to the east that are poorly known and rarely visited.

Biogeographical zones

(based on *The Indicative Map of European Biogeographical Regions: Methodology and development*. European Topic Centre on Biological Diversity. Muséum National d'Histoire Naturelle. 2006)

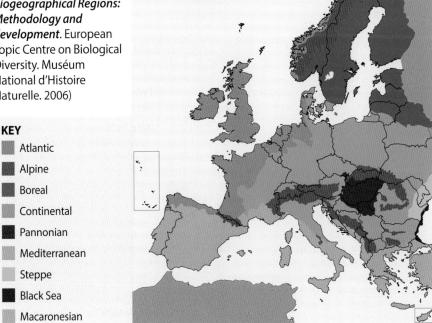

KEY

- Atlantic
- Alpine
- Boreal
- Continental
- Pannonian
- Mediterranean
- Steppe
- Black Sea
- Macaronesian

The types of dragonfly

The 140 species of dragonfly breeding in Europe fall into seven broad groups ('types') of damselfly (Zygoptera) and nine types of dragonfly (Anisoptera). This page-spread (covering damselflies) and the following (covering dragonflies) show representative males of each type to illustrate typical appearance and structure. Further details about the types can be found on *pp. 16–17* (damselflies) and *pp. 122–125* (dragonflies), where links are provided to the relevant section and individual species accounts. Females and immatures are similar in shape and structure to mature males, but differ in their colours and patterning. The images shown here are broadly to scale and at about life-size, reflecting the average size of all the species within their group.

DAMSELFLIES (ZYGOPTERA)

Key features of an adult damselfly

- ◆ Typically small size and dainty proportions
- ◆ Flight usually weak and brief
- ◆ Wings usually held together over back
- ◆ Front and hind wings identical in shape
- ◆ Eyes separated
- ◆ Lay eggs into plants above or below water

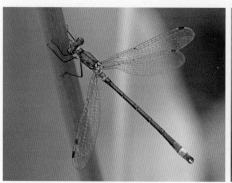

Spreadwings (*p. 18*)

Winter damsels (*p. 18*)

Demoiselles (*p. 38*)

Odalisque (*p. 48*)

Red damsels (*p. 50*)

Featherlegs (*p. 58*)

Bluets (*p. 64*)

Brighteyes (*p. 94*)

Bluetails (*p. 94*)

Sedgling (*p. 94*)

DRAGONFLIES (ANISOPTERA)

Key features of an adult dragonfly

◆ Robust form
◆ Flight powerful, often persistent and rapid
◆ Wings held at right angles to body at rest

◆ Front and hind wings have different shapes
◆ Eyes meet (except in clubtails, *etc.*)
◆ Lay eggs into plants, debris or water

Hawkers & spectres (*p. 126*)

Emperors (*p. 126*)

Clubtails, snaketail, pincertails, hooktail &
Bladetail (*p. 176*)

Goldenrings, cruiser & cascader (*p. 208*)

Emeralds (*p. 230*)

Baskettail & chasers (*p. 250*)

Skimmers (*p. 258*)

Whitefaces (*p. 282*)

Darters, dropwings, pennant, percher &
groundling (*p. 294*)

Gliders (*p. 338*)

How to identify dragonflies

The important features to focus on are shown in the illustrations opposite. The most useful features to check when identifying dragonflies are summarized in the following table:

Damselflies	Head / Thorax / Abdomen / Wings / Legs	Dragonflies
Eye colour	Head	Eye colour; frons ('face') colour and pattern
Presence and colour of stripes on the top and side; shape of pronotum	Thorax	Presence, extent and colour of stripes on the top and side
Colour; markings on top, especially of the second, and eighth to tenth abdominal segments (abbreviated as S2 and S8–10)	Abdomen	Colour; markings on top, especially of the second, and eighth to tenth abdominal segments (abbreviated as S2 and S8–10); presence of a 'waist'
Colour and shape of wing-spots	Wings	Colour at base, and of veins (especially the costa); colour and shape of wing-spots
Colour	Legs	Colour(s)

It is very useful to gain a good knowledge of the commoner dragonflies as a starting point from which to identify similar species. Remember that immatures have subdued colours (dull eye colour, pale wing-spots and rather reflective wings), and a rather weak, unsteady flight. If it is possible to obtain close views, it is often worth sexing the individual: look on the underside near the front of the abdomen for the bulge of the male's accessory genitalia, and near the tip for the female's ovipositor. Try to judge size only when an individual can be compared directly with another. It is helpful to learn the flight and perching characteristics of different species, as well as their habitat preferences or tolerances (*e.g.* acidic, enriched, brackish, standing or running water). The distribution maps and flight periods given in each species account may help in the identification of some species.

Chrysopa perla
Lacewing (Neuroptera)

Macronemurus appendiculatus
Antlion (Neuroptera)

Possible confusion groups
Some other insect groups that might be mistaken for dragonflies are illustrated on this page; all have conspicuous antennae and/or a long 'tail', unlike dragonflies.

Ephemera vulgata
Mayfly (Ephemeroptera)

Libelloides coccajus
Owl-fly [Ascalaphid] (Neuroptera)

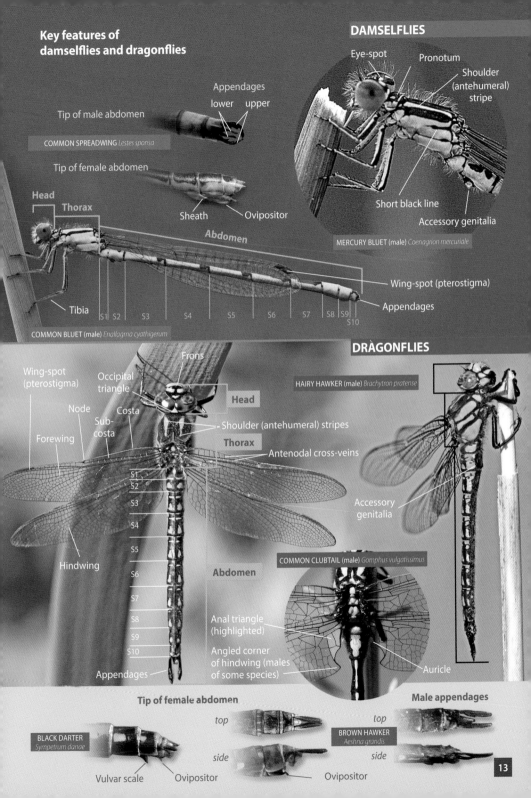

Key features of damselflies and dragonflies

DAMSELFLIES

Eye-spot
Pronotum
Shoulder (antehumeral) stripe

Appendages
lower   upper

Tip of male abdomen

COMMON SPREADWING *Lestes sponsa*

Short black line

Accessory genitalia

MERCURY BLUET (male) *Coenagrion mercuriale*

Tip of female abdomen

Head

Thorax

Sheath   Ovipositor

Abdomen

Wing-spot (pterostigma)

Appendages

Tibia   S1 S2   S3   S4   S5   S6   S7   S8 S9 S10

COMMON BLUET (male) *Enallagma cyathigerum*

DRAGONFLIES

Frons

Wing-spot (pterostigma)

Occipital triangle

Head

HAIRY HAWKER (male) *Brachytron pratense*

Node

Costa

Shoulder (antehumeral) stripes

Sub-costa

Thorax

Forewing

Antenodal cross-veins

S1
S2
S3
S4
S5
S6
S7
S8
S9
S10

Accessory genitalia

COMMON CLUBTAIL (male) *Gomphus vulgatissimus*

Hindwing

Abdomen

Anal triangle (highlighted)

Angled corner of hindwing (males of some species)

Auricle

Appendages

Tip of female abdomen

Male appendages

top

top

BLACK DARTER
Sympetrum danae

BROWN HAWKER
Aeshna grandis

side

side

Vulvar scale   Ovipositor

Ovipositor

13

Glossary

accessory genitalia *	structure on underside of **S2** of male's abdomen that holds sperm prior to copulation (also known as secondary genitalia)
anal triangle *	clear triangle of cells at rear of base of hindwing
antenodal cross-veins *	veins at right angles to the **costa**, between the body and **node**
appendages *	2–4 extensions at tip of the abdomen, in males used to clasp the female during copulation
auricle *	small protuberances at side of **S2** in males of some Anisoptera
costa *	thick front vein that forms leading edge of wing
endemic	native, with range restricted to a defined area (country or region)
eye-spots *	pale spots behind the eyes of some damselflies, technically known as post-ocular spots
frons *	uppermost part of front of head ('face')
immature	an adult that has not yet attained the full colouring typically associated with sexual maturity
in tandem	when male and female are joined (before or after copulation)
iridescent	showing colours (typically green, blue or bronzy) that change with the angle of observation and/or illumination
node (or nodus) *	notch halfway along leading edge of wing
obelisk position	posture adopted by some dragonflies in which the abdomen is pointed towards the sun to avoid overheating
occipital triangle *	triangular area behind eyes
ovipositor *	blade-like egg-laying structure under **S9–10** of female dragonflies that deposits eggs into plant material or substrates
pronotum *	shield-like plate that covers top of front of thorax
pruinose	coated in a waxy bloom, often producing a pale blue colour (in this book, referred to as 'pruinescence')
S2, S3, S8–10, *etc.* *	abbreviations used for the number, or range, of abdominal segment(s), counted from the base
shoulder stripes *	pairs of pale (antehumeral) or dark (humeral) stripes on top edges of thorax
spp. / ssp.	abbreviations for species (plural) / subspecies (singular)
sub-costa *	next main vein back from the **costa**
teneral	newly-emerged adult dragonfly, with soft cuticle and shiny wings, lacking full adult coloration
tibia(e) *	long leg segment below the 'knee', equivalent to 'shin' in humans
venation	network of veins in the wings
vulvar scale *	flap below **S8–9**, present in dragonflies that lay eggs freely onto the water surface; prominent in some species; not present in damselflies
wing-spot *	dark or coloured cell along leading edge of wing, towards the tip, technically known as pterostigma (female demoiselles have a whitish pseudopterostigma or 'false wing-spot')

* term illustrated on previous page

All 140 species recorded in Europe or W Turkey are afforded a full account, grouped as follows:

Pages 16–121: **damselflies (Zygoptera) [47 species]**;

Pages 122–341: **dragonflies (Anisoptera) [93 species]**.

The species accounts follow broadly the generally accepted taxonomic order of families, but within each section (which relate to the broad 'types' defined earlier) similar species are grouped together for ease of comparison. The text for each is presented as follows:

CONSERVATION STATUS
For details see *p. 342*.

Species is endemic to Europe

NOMENCLATURE The English & *scientific* names used follow Dijkstra & Lewington (2006).

 English name *Scientific name* ×2

LEGISLATION Protected by European legislation.
For details see *p. 342*.

Alternative English language names used in Europe

Summary paragraph, including comments on the species' status.

GENERAL STATUS
Summary of overall status across Europe plus population trend, where:

= – stable
▲ – increasing
▼ – declining
❓ – unknown

PROTECTED

Status [trend]

Main breeding habitat(s)

Identification:
A concise description of the adult forms, detailing the key identification features, with the most important highlighted in **bold** text that, in combination, are diagnostic.

Behaviour: Aspects of behaviour that are a clue to identification.

SCALE BAR and SCALE
Bar shows actual maximum and minimum total length. The scale of all photos shown on the accompanying plate is given next to *Scientific name*.

Breeding habitat: A brief summary of the species' habitat preferences.

DISTRIBUTION MAP
■ = recorded since 1990
● = isolated record

J F M A M J J A S O N D

FLIGHT PERIOD
The darker the shade of green, the more likely to be seen during that time of year

Overall length: xx–xx mm
Hindwing: xx–xx mm

AREA MODIFIERS
regions of areas/countries are abbreviated as follows:
C (central), E (east),
N (north), NE (north-east),
NW (north-west), S (south),
SE (south-east), SW (south-west), W (west).

MEASUREMENTS
Typical adult biometrics

LOOK-ALIKES (in range)
Species/forms likely to cause confusion in range.

THE PLATES
Typically showing both sexes and immature forms from various angles to show key features, scaled optimally for each group. The key identification features to look for are highlighted.

The following annotations are used: male female immature old

Introduction to damselflies (Zygoptera)

Damselflies are typically smaller than dragonflies, with a weaker structure and less powerful and sustained flight. All four wings are identical and are usually held together over the abdomen when at rest, partially spread in some species and, in one case (Odalisque), held straight out in the manner of a dragonfly. The eyes are positioned on the sides of the head and well separated. Most species spend much time near wetland margins, while a few habitually fly out or perch over water away from the water's edge. All lay eggs into plant material, either living or dead and sometimes woody, above, at or below the surface of the water. Despite their relatively fragile appearance, some move long distances and may be found regularly away from water.

This section summarizes the key features of the 'types' of damselfly recorded in Europe. The 47 species of damselfly fall into seven broad groups, two of which are further subdivided. A male from each group or sub-group is shown here.

SPREADWINGS & WINTER DAMSELS   *(p. 18)*
GENERA: *Lestes, Chalcolestes* & *Sympecma*   9 species

Spreadwings (emerald damselflies)   *(p. 18)*
GENERA: *Lestes* & *Chalcolestes*   7 species

Medium-sized; metallic green; typically rest with wings held open at about 60° to abdomen; some species are pruinose; standing waters, including temporary wetlands.

Spreadwing

Winter damsels   *(p. 19)*
GENUS: *Sympecma*   2 species

Medium-sized; brownish with metallic 'rocket'-shaped markings on abdomen; wing-spots on forewing are nearer the wingtip than those on hindwing; standing waters.

Winter damsel

DEMOISELLES (or jewelwings)   *(p. 38)*
GENUS: *Calopteryx*   4 species

Unmistakable large damselflies; metallic green, blue or copper-coloured bodies; dark wings; distinctive fluttering wingbeats; flowing waters.

Demoiselle

ODALISQUE   *(p. 48)*
GENUS: *Epallage*   1 species

Large; uniquely holds wings out from body like dragonfly; ephemeral flowing waters in SE Europe.

Odalisque

RED DAMSELS (p. 50)

GENERA: *Pyrrhosoma* & *Ceriagrion*   4 species

Small to medium-sized; at least some red on abdomen; one or two black lines across side of thorax; mainly standing waters.

Red damsel

FEATHERLEGS (p. 58)

GENUS: *Platycnemis*   3 species

Medium-sized; body pale blue, whitish or orange; additional pair of black shoulder stripes; wing-spots rectangular, orange or brown; tibiae broadened, feather-like; mainly flowing waters.

Featherleg

BLUETS (p. 64)

GENERA: *Enallagma* & *Coenagrion*   13 species

Medium-sized; black-and-blue; pale shoulder stripes; wing-spots usually black, more-or-less diamond-shaped; standing or flowing waters.

Bluet

'BLUE-TAILED' DAMSELFLIES (p. 94)

Erythromma, Ischnura & *Nehalennia*   13 species

Brighteyes (p. 94)

GENUS: *Erythromma*   3 species

Medium-sized; males have blue on S9–10 and red or blue eyes; females have largely black top of abdomen; wing-spots dark or pale; standing or slow-flowing waters (males typically sit on floating or low, emerging vegetation).

Brighteye

Bluetails (p. 94)

GENUS: *Ischnura*   9 species

Small or very small; dark abdomen with blue S8 or S8–9 (may be absent on females); wing-spots bicoloured; standing or slow-flowing waters.

Bluetail

Sedgling (p. 94)

GENUS: *Nehalennia*   1 species

Very small; metallic green; male has blue on S8–10; pale wing-spots; inconspicuous; standing waters.

Sedgling

SPREADWINGS & WINTER DAMSELS

FAMILY | **Lestidae**

GENERA | *Lestes, Chalcolestes & Sympecma*

9 SPECIES

Medium to large damselflies (mean lengths 37–47 mm) with clear, narrow-based wings held partially spread (spreadwings) or together (winter damsels). Large, rectangular, brown or bicoloured wing-spots. Body shiny metallic green or bronzy, with or without pale blue pruinescence. Male appendages are relatively large and pincer-like. The ovipositor is large; eggs are laid into plant material, mostly in tandem and above water. Breed in mostly standing, sometimes temporary, waters.

Key features of spreadwings

- ◆ wings typically well spread
- ◆ relatively large damselflies
- ◆ plain metallic green or bronzy; some species pruinose
- ◆ ♂ appendages / ♀ ovipositor prominent

Identify species by markings on side of thorax and back of head | extent of pruinescence | wing-spot colour | shape of male appendages / female ovipositor

Spreadwings

GENERA: *Lestes* (pp. 20–28) & *Chalcolestes* (pp. 30–33), also known as emerald damselflies (7 species), are relatively large, metallic green damselflies, becoming more bronzy with age. They rest typically with their wings spread at about 60° to the abdomen. Wing-spots are rectangular and relatively large, abutting two or three cells. The males (and some females) of most species become pruinose. Males have conspicuous, calliper-like appendages, while females have a robust ovipositor that is used to insert eggs into plant stems or even twigs, above and sometimes away from water. Eggs overwinter and larvae develop rapidly, with adult numbers often peaking in mid- or late summer. All spreadwings breed in standing waters, often well-vegetated, although the willow spreadwings (*Chalcolestes*) also breed in flowing water. Standing water sites often have fluctuating water levels and may even dry up completely, which probably explains why adults may disperse far from known breeding sites. Some species have expanded their range northwards in recent decades. Adults can be inconspicuous, keeping within dense vegetation or, in the case of willow spreadwings, hanging below tree branches, often in the shade (willow spreadwings are the only European Dragonflies to lay eggs into living woody stems).

COMMON SPREADWING
Lestes sponsa

Egg-laying willow spreadwings use serrations ('teeth') on the underside of the ovipositor to cut into bark during egg-laying, leaving distinctive scars that may be seen for several years.

WESTERN WILLOW EMERALD
Chalcolestes viridis

Spreadwings compared

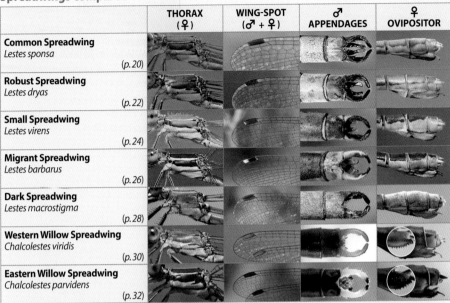

	THORAX (♀)	WING-SPOT (♂ + ♀)	♂ APPENDAGES	♀ OVIPOSITOR
Common Spreadwing *Lestes sponsa* (p. 20)				
Robust Spreadwing *Lestes dryas* (p. 22)				
Small Spreadwing *Lestes virens* (p. 24)				
Migrant Spreadwing *Lestes barbarus* (p. 26)				
Dark Spreadwing *Lestes macrostigma* (p. 28)				
Western Willow Spreadwing *Chalcolestes viridis* (p. 30)				
Eastern Willow Spreadwing *Chalcolestes parvidens* (p. 32)				

Winter damsels GENUS: *Sympecma* (pp. 34–37)

(2 species) are medium-sized (around 37 mm) and cryptically coloured, with clear wings held together over the abdomen like most other damselflies, but habitually both to one side. Unusually, the large, brown wing-spot on the forewing is closer to the wingtip than that on the hindwing. Upper surfaces of the body are metallic bronzy or green, S3–7 having 'torpedo'-shaped markings. Male appendages are pale, pincer-like on males, large and pointed on females. Identification of species is by subtle differences in thorax pattern and male appendages. Uniquely in Europe, adults are present year-round, emerging from June and overwintering in rough, typically wooded, habitats; they return in spring to well-vegetated standing waters, where eggs are laid in tandem, often into floating dead plant material.

Key features of winter damsels

◆ wings typically held closed and to one side of abdomen

◆ relatively large damselflies

◆ dark, slightly metallic with 'torpedo'-shaped markings on abdomen

◆ wing-spot on forewing closer to wingtip than that on hindwing

◆ ♂ appendages prominent

Identify species by thorax markings | length and shape of male appendages

COMMON WINTER DAMSEL ♂
Sympecma fusca

Winter damsels compared

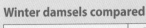

	THORAX	♂ APPENDAGES
Common Winter Damsel *Sympecma fusca* (p. 34)		
Siberian Winter Damsel *Sympecma paedisca* (p. 36)		

LC Common Spreadwing

Lestes sponsa ×2

Emerald Damselfly

| Common | = |

| Various standing waters |

J F M A M J J A S O N D

| **Overall length:** | 35–39 mm |
| **Hindwing:** | 19–24 mm |

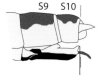

S9   S10

♀ ovipositor just reaches
tip of S10

The most widespread spreadwing, although scarcer in the S of its range, found at shallow, well-vegetated standing waters. As with most other spreadwings, numbers peak in late summer, later than most other damselflies.

Identification: Rather large damselfly with wings typically held well-spread at rest. Very similar to Robust Spreadwing (*p. 22*) but relatively slender. Body metallic bronzy-green; wing-spots develop through brown to blackish. ♂ Powder blue pruinescence develops on S1–2, S9–10 (plus S8 on some) as well as on the pronotum and side of thorax; eyes blue; appendages prominent, the smaller, lower pair **straight**. ♀ Lower lobes to prothorax pale; isolated metallic spot just above base of middle leg; thin, yellowish shoulder stripes; abdomen distinctly thicker than male, lacking any blue and with beige underparts; pair of **rounded** dark spots on S1; ovipositor prominent, just **reaching** tip of S10; eyes brown. VARIATION: Immatures have underside and pruinose areas pinkish-buff; body becomes bronzy on old individuals; females rarely pruinose.

Behaviour: Flies rather weakly and tends to favour tall, dense vegetation, perching rather inconspicuously. Matures relatively slowly, after two weeks or so.

Breeding habitat: Favours smaller, shallow standing waters (*e.g.* bog pools, ponds and ditches) with luxuriant tall grass, rushes (*Juncus* spp.) or sedges (*Carex* spp.); some sites dry out in late summer.

LOOK-ALIKES (in range)
Other spreadwings (*pp. 22–33*)

♀

♀

♂i

♀i

♀ rounded front edge to S1 marks

♂

♂

♂ S2 all-blue; lower appendages straight

LC Robust Spreadwing

Lestes dryas ×2

Scarce Emerald Damselfly, Turlough Spreadwing

| Fairly common | = |
| Pond, lake, ditch | |

J F M A M J J A S O N D

| **Overall length:** | 35–40 mm |
| **Hindwing:** | 20–27 mm |

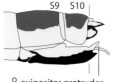

S9   S10

♀ ovipositor protrudes
beyond tip of S10

As its name suggests, a rather robust spreadwing that inhabits shallow, sometimes temporary waters where few other dragonflies can survive. More common in S of range.

Identification: Rather large and robust damselfly with wings typically held well-spread at rest. Body largely metallic green with bronzy iridescence, **more robustly built** than Common Spreadwing (*p. 20*) with slightly **broader wing-spots**. ♂ Powder blue pruinescence on S1, usually only **front two-thirds of S2**, S9–10 (plus S8 on some) as well as on the pronotum and side of thorax; eyes blue (brighter than on Common Spreadwing); lower appendages distinctively thick and **incurved** at tip. ♀ Thorax and abdomen rather broad, lacking any blue and with greenish-cream underparts; lower lobes to pronotum metallic green; extensive dark metallic area above base of middle leg; very thin, incomplete pale shoulder stripes; pair of **squarish**, dark marks on S1; ovipositor prominent, just **protruding** beyond tip of S10; eyes brown. VARIATION: Immatures metallic chocolate-emerald with greenish-cream underparts (Common Spreadwing pinkish-buff); body becomes bronzy on old individuals; female rarely pruinose.

Behaviour: Like other spreadwings, has a rather weak flight and is usually found in tall, dense vegetation, perching rather inconspicuously.

Breeding habitat: Shallow, well-vegetated ponds, lakes and ditches, often drying out to some extent.

LOOK-ALIKES (in range)
Other spreadwings (*pp. 20–33*)

♀i

♀

♀ some pruinose

♀

♂

♀

♀ straight front edge to S1 marks

♂

♂ S2 ⅔ blue; lower appendages incurved

LC Small Spreadwing

Lestes virens ×2

Small Emerald Damselfly

Locally common =

Pond, lake, marsh

NORTH
SOUTH
J F M A M J J A S O N D

Overall length:	30–39 mm
Hindwing:	19–23 mm

True to its name, this species is, on average, the smallest of the spreadwings, and occurs in richly vegetated wetlands, often with other *Lestes* species.

Identification: Similar to other spreadwings, sharing habit of holding wings partially spread, but smaller and with metallic green upperparts. **Back of head yellow** (as Migrant Spreadwing (*p. 26*)), contrasting sharply with green top of head; pale shoulder stripes narrow, not extending back to base of forewings; wing-spots brown with **white** sides. ♂ Eyes blue; blue **pruinescence limited to S9–10**, around wing bases and, when fully mature, under thorax and on pale shoulder stripes; lower appendages **short and straight**. ♀ Ovipositor short, like Common Spreadwing (*p. 20*), but pale; sheath slightly pointed and pale. VARIATION: In SW Europe, slightly less extensive dark top to thorax and pale shoulder stripes extend to base of forewings (named ssp. *virens*, but taxonomy unclear); in W Iberia, wing-spots bicoloured. Old individuals darken and may become pruinose on underside of thorax and tip of abdomen.

Behaviour: Favours dense, rushy vegetation, where perches rather inconspicuously; although appears to be a rather weak flier, has a tendency to wander to drier grassy or heathy areas.

Breeding habitat: Richly vegetated ponds, lakes, marshes and, in NW of range, peaty sites with bog-mosses (*Sphagnum* spp.) and rushes (*Juncus* spp.); these include seasonally wet and brackish sites and are often close to woodland.

LOOK-ALIKES (in range)
Other spreadwings (*pp. 20–33*)

♂ lacks blue at base of abdomen when mature

♂ + ♀ back of head yellow

♂ + ♀ wing-spots with obvious white sides

♂ lower appendages short and straight

LC Migrant Spreadwing

Lestes barbarus ×2

Southern Emerald Damselfly

| Common | = |

| Seasonally flooded standing waters |

NORTH
SOUTH
J F M A M J J A S O N D

| **Overall length:** | 40–45 mm |
| **Hindwing:** | 20–27 mm |

A large spreadwing, easily distinguished by its brown-and-white wing-spots.

Identification: Large damselfly with wings typically held well-spread at rest; differs from other spreadwings in having long, obviously **bicoloured** wing-spots, generally pale appearance and clear **yellowish shoulder stripes**. Both sexes mainly metallic green, with contrasting yellow or pale green side of body, extending to top on S9–10; eyes greenish; rear of head **yellow or greenish**, contrasting sharply with dark top of head. ♂ Generally lacks blue pruinescence; S10 is whitish; from above, upper appendages pale with dark tips, lower **small, pointed and outcurved** at tips. ♀ Ovipositor usually **pale**. VARIATION: As with other spreadwings, wing-spots wholly pale on immatures and body becomes bronzy on old individuals.

Behaviour: Breeding biology similar to other spreadwings, but eggs may be laid into woody stems, such as those of willows (*Salix* spp.), as well as rushes (*Juncus* spp.), sedges (*Carex* spp.) and other wetland plants.

Breeding habitat: Specialist of various seasonally flooded standing waters that typically dry out early in summer; eggs often laid in apparently dry areas; prone to wandering and often found well away from water. Absent from N Europe, but has spread N in recent decades.

LOOK-ALIKES (in range)
Other spreadwings (*pp. 20–33*)

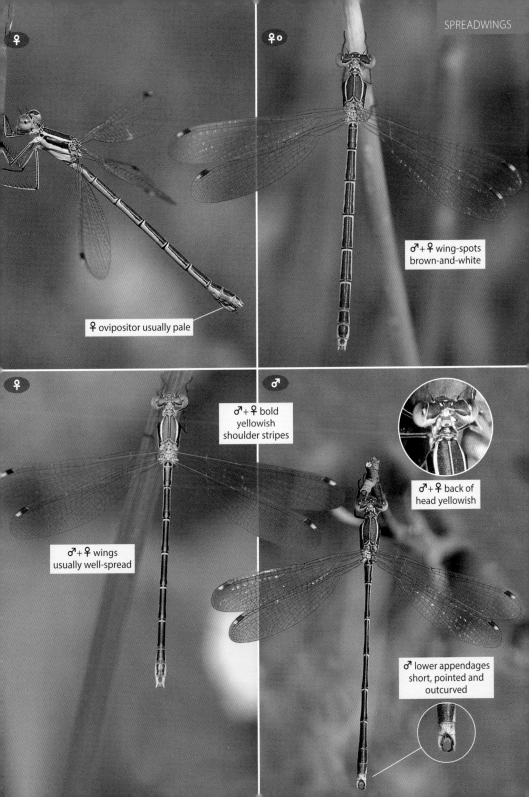

♀ ovipositor usually pale

♂ + ♀ wing-spots brown-and-white

♂ + ♀ bold yellowish shoulder stripes

♂ + ♀ back of head yellowish

♂ + ♀ wings usually well-spread

♂ lower appendages short, pointed and outcurved

VU Dark Spreadwing

Lestes macrostigma ×2

Dark Emerald Damselfly

Very locally common ▽

Brackish standing waters

J F M A M J J A S O N D

| **Overall length:** | 39–48 mm |
| **Hindwing:** | 24–27 mm |

A large, early emerging spreadwing that occurs at a highly fragmented range of brackish sites in S Europe, reflecting the scarcity of its preferred habitat. Both sexes develop pruinescence and are more alike than in other spreadwings.

Identification: Large damselfly with wings typically held well-spread at rest; when mature, both sexes lack overall metallic appearance of other spreadwings, except perhaps on the mid-section of abdomen; they rapidly develop **extensively blue-purple** pruinescence; wing-spots black and **large**, bordered by 3–4 cells along rear edge. ♂ Abdomen metallic bronzy-green on S3–7, otherwise entire body pruinose pale blue; top of thorax may be darker purplish; eyes dark blue or purple. ♀ Like male but slightly less pruinose, especially on top of thorax; ovipositor dark and pruinose. VARIATION: Immatures metallic green and yellow, soon pruinose.

Behaviour: Adults emerge earlier in the season than other spreadwings, and mostly within a 1–2-week period. Most are found at or close to suitable breeding habitat, where can be locally abundant; individuals occasionally turn up some distance away, suggesting a tendency to disperse.

Breeding habitat: Typically shallow, stagnant, often brackish waters dominated by club-rushes (*Bolboschoenus* spp.). Found mainly at coastal saltmarsh ponds, saline lagoons and derelict saltpans, but also inland at steppe lakes and small reservoirs. Sites may dry out during summer.

LOOK-ALIKES (in range)

Other spreadwings (*pp. 20–33*)

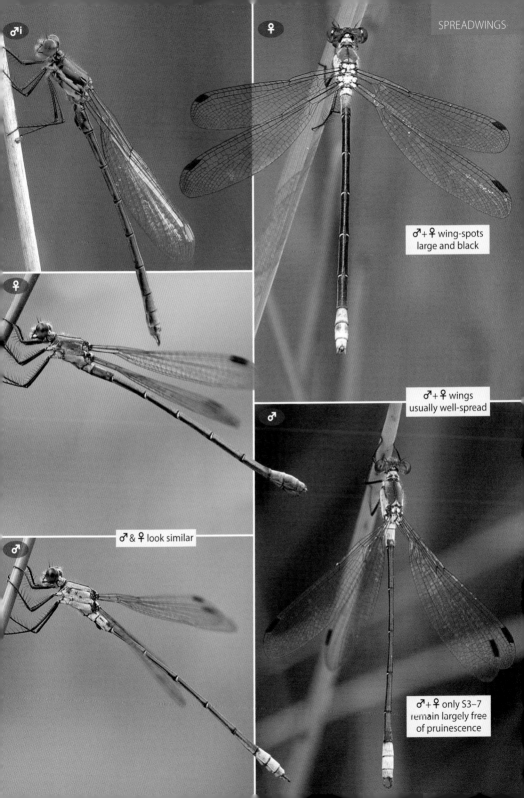

♂i

♀

♂ + ♀ wing-spots
large and black

♀

♂ + ♀ wings
usually well-spread

♂ & ♀ look similar

♂

♂

♂ + ♀ only S3–7
remain largely free
of pruinescence

LC **Western Willow Spreadwing**

Chalcolestes viridis ×2

Willow Emerald Damselfly

Common ═ (recent spread N)

Lake, pond, river, stream

J F M A M J J A S O N D

Overall length:	39–48 mm
Hindwing:	23–28 mm

♀ ovipositor has >9 'teeth'

A large, cryptically coloured spreadwing that is easily overlooked due to its association with waterside trees. Extremely similar to Eastern Willow Spreadwing (*p. 32*) and specific identification requires careful examination (ideally of several individuals). Willow spreadwings with intermediate characters may occur in Italy and N Balkans, where the ranges of the two species overlap.

Identification: Larger and darker than other spreadwings, without blue eyes or pruinescence; both sexes mainly metallic green or bronzy; wing-spots large and **pale**, outlined in black (beware immature *Lestes* spreadwings, which also have pale wing-spots and lack pruinescence); thorax has thin yellow shoulder stripes, lower border of metallic green on side irregular forming a **dark 'spur'** (may be present in reduced form on *Lestes* spreadwings). ♂Upper appendages more than 3× length of lower, distinctly whitish, with **black extending from tip at least along outer edge** and 'tooth' near tip of inner edge relatively **blunt**; lower appendages short and **straight**. ♀ Ovipositor dark with pale area on lower edge and **more than nine 'teeth'** on lower edge at tip. VARIATION: Body becomes bronzy on old individuals.

Behaviour: Hangs with wings spread wide, often in shade of trees near breeding waters. Eggs are laid into twigs (sometimes non-woody plants) overhanging water using serrations on underside of ovipositor, leaving pairs of oval scars in bark, which may be visible for several years.

Breeding habitat: Permanent standing and flowing waters with marginal and overhanging trees and bushes.

LOOK-ALIKES (in range)

Other spreadwings (*pp. 20–33*)

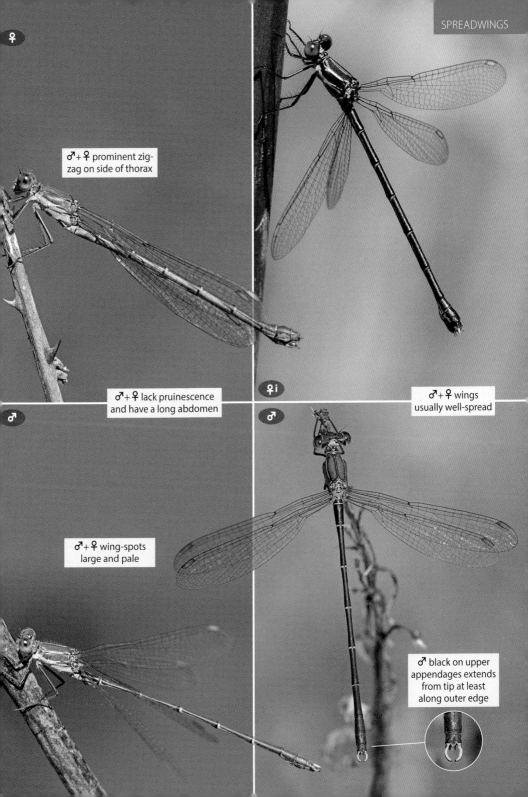

♀

♂ + ♀ prominent zig-zag on side of thorax

♂ + ♀ lack pruinescence and have a long abdomen

♀ i

♂ + ♀ wings usually well-spread

♂

♂ + ♀ wing-spots large and pale

♂

♂ black on upper appendages extends from tip at least along outer edge

LC Eastern Willow Spreadwing

Chalcolestes parvidens ×2

Locally common ?

Lake, pond, river, stream

J F M A M J J A S O N D

Overall length:	44–50 mm
Hindwing:	22–26 mm

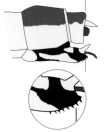

♀ ovipositor has <9 'teeth'

The south-eastern member of a very similar pair of large, cryptically coloured spreadwings, formerly treated as a subspecies of Western Willow Spreadwing (*p. 30*). The two species are extremely similar and identification is best confirmed in the hand (ideally check several individuals). Willow spreadwings with intermediate characters may occur in Italy and N Balkans, where the ranges of the two species overlap.

Identification: Larger and darker than other spreadwings, and lacking blue eyes or pruinescence; both sexes mainly metallic green or bronzy; wing-spots large and brown (darker than mature Western Willow Spreadwing); thorax has thin yellow shoulder stripes, lower border of metallic green on side irregular forming a **dark 'spur'** (may be present in reduced form on *Lestes* spreadwings). ♂ Upper appendages more than 3× length of lower, distinctly **whitish, with black only at extreme tip** and 'tooth' near tip of inner edge relatively **pointed**; lower appendages short, with **upturned tip**. ♀ Ovipositor has **fewer than nine 'teeth'** on lower edge at tip. VARIATION: Body becomes bronzy on old individuals.

Behaviour: Hangs with wings spread wide, often in the shade of trees near breeding waters, for much of the summer. Eggs laid into twigs (sometimes non-woody plants) overhanging water, leaving distinctive scars in bark. Possibly flies earlier in the year than Western Willow Spreadwing and may be more active in mornings, favouring shade in hot, sunny conditions.

Breeding habitat: Permanent standing and flowing waters with marginal and overhanging trees and bushes.

LOOK-ALIKES (in range)

Other spreadwings (*pp. 20–31*)

♂ + ♀ prominent zig-zag on side of thorax

♂ + ♀ wing-spots large and brown

♀

♀

♂ + ♀ lack pruinescence and have a long abdomen

♂ + ♀ wings usually well-spread

♂

♂ black on upper appendages only at extreme tip

LC Common Winter Damsel

Sympecma fusca ×2

Winter Damselfly

Common △ (expanding N)

Various standing waters

J F M A M J J A S O N D

Overall length:	34–39 mm
Hindwing:	18–23 mm

S10

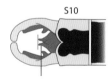

♂ lower appendages at least reach 'teeth' on upper appendages

Although as large as some spreadwings, the lack of bright colours on the winter damsels means that they can easily be overlooked. Widespread, except in N Europe; there is some overlap with the very similar Siberian Winter Damsel (*p. 36*) in C and E Europe.

Identification: Medium-sized; pale fawn (but darker after winter), with dark, slightly metallic markings on top of thorax and abdomen. Metallic colouring recalls spreadwings (*pp. 20–33*), but wings are held closed when at rest and habitually to one side of abdomen. Wing-spots long, brown; wing-spot on forewing nearer wingtip than on hindwing. Blue spots on top of eyes in spring. Combination of broad, brownish shoulder stripes and 'torpedo'-shaped dark markings on S3–6 resembles dull female Common Bluet (*p. 68*). Told from Siberian Winter Damsel by **straight division** between pale shoulder stripe and dark area above, and by continuous dark line below; otherwise only differs in small details of appendages. ♂ Upper appendages pale and pincer-shaped (as in spreadwings and Siberian Winter Damsel); lower appendages **reach or extend beyond** 'teeth' on upper appendages. ♀ Appendages longer than those of spreadwings and straight, but ovipositor shorter; only distinguishable from Siberian Winter Damsel by subtle differences in thoracic markings.

Behaviour: Often seen away from water, along sheltered rides in woodland. Inconspicuous in colour and behaviour, typically sitting with abdomen tight against perch. Winter damsels are the only European dragonflies to overwinter as adults; they can be seen year-round. Tandem pairs lay eggs into floating remains of plants such as rushes (*Juncus* spp.).

Breeding habitat: Fairly shallow, mainly lowland, standing waters, sometimes slow-flowing or brackish, with abundant emergent vegetation. In autumn, moves to sheltered areas, where overwinters among dry grasses or twigs, or in woodpiles, even on exposed perches, occasionally rousing to fly in warm weather.

LOOK-ALIKES (in range)
Siberian Winter Damsel (*p. 36*)
♀ Common Bluet (*p. 68*)

34

♀ July

♂ + ♀ wings held together, to one side of abdomen

♂

♂

♀ September

♂ + ♀ wing-spot nearer wingtip on forewing

♂

♂ + ♀ straight upper margin to pale shoulder stripe

♂ + ♀ dark line below pale shoulder stripe broad

♀ March

♀ blue eyes and dark thorax in spring

♂

LC Siberian Winter Damsel

Sympecma paedisca ×2

PROTECTED

Locally common ▽

Various standing waters

J F M A M J J A S O N D

Overall length:	36–39 mm
Hindwing:	18–22 mm

S10

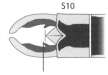

♂ lower appendages do not reach 'teeth' on upper appendages

North-eastern equivalent of almost identical Common Winter Damsel (*p. 34*), differing only in minor details of thorax pattern and appendages. Most common in E Europe but rare and very local where the ranges of the two species overlap in C Europe.

Identification: Medium-sized and, as Common Winter Damsel, pale fawn (but darker after winter), with dark, slightly metallic markings on top of thorax and abdomen. Metallic colouring recalls spreadwings (*pp. 20–33*), but wings are held closed when at rest and habitually to one side of abdomen. Wing-spots long, brown; wing-spot on forewing nearer wingtip than on hindwing. Blue spots on top of eyes in spring. Combination of broad, brownish shoulder stripes and 'torpedo'-shaped dark markings on S3–6 resembles dull female Common Bluet (*p. 68*). Differs from Common Winter Damsel in the lower dark line across side of thorax usually being **thinner (discontinuous in some)** and with a small, dark **'bulge'** protruding into the top of the pale shoulder stripe; otherwise differs only in small details of appendages.
♂ Upper appendages pale and pincer-shaped, as in spreadwings and Common Winter Damsel; lower appendages very short and **do not reach** 'teeth' on base of inner edge of upper appendages.
♀ Appendages longer than those of spreadwings and straight, but ovipositor shorter; only distinguishable from Common Winter Damsel by subtle differences in thoracic markings.

Behaviour: Often seen away from water, along sheltered rides in woodland. Inconspicuous in colour and behaviour, typically sitting with abdomen tight against perch. Adults overwinter and can be seen year-round. Tandem pairs lay eggs into floating remains of plants such as rushes (*Juncus* spp.).

Breeding habitat: As with Common Winter Damsel, occupies a wide range of standing waters; in Alps found in heathland and peat bogs.

LOOK-ALIKES (in range)
Common Winter Damsel (*p. 34*)
♀ Common Bluet (*p. 68*)

♀

♂

♂ + ♀ wings held together, to one side of abdomen

♀

♂

♂ + ♀ wing-spot nearer wingtip on forewing

♂ + ♀ dark 'bulge' protrudes into top of pale shoulder stripe

♀

♂

♂ + ♀ dark line below pale shoulder stripe narrow (discontinuous on some individuals)

DEMOISELLES

GENUS | *Calopteryx*

FAMILY | **Calopterygidae**

4 SPECIES

Unmistakable very large damselflies (mean length 47 mm) with metallic green, blue or copper-coloured bodies, broad, pigmented wings and distinctive fluttering wingbeats. Eggs are laid into plant material, initially in tandem at the water's surface but then the female alone submerges and continues laying. Breed in flowing waters.

Key features of demoiselles

◆ very large damselflies
◆ pigmented wings
◆ fluttering flight
◆ metallic body
◆ breed in flowing waters

Identify species by wing and body colour | extent of dark in wing

Demoiselles (also known as jewelwings) are distinctive, very large damselflies with fluttering wingbeats. They have a dark, metallic body, the colour of which ranges from blue or green, through dark reddish-purple to almost black. The wings are broad and lack the narrow, 'stalked' bases of most other damselflies, tapering evenly to the base. In all species the wings are tinted or have dark patches, and there is geographical variation in the extent of dark pigmentation. Males lack wing-spots but females have whitish 'false wing-spots' (see *below*). The legs are long and bristly. The abdomen is typically raised towards the tip when perched. All species breed in flowing water, where large numbers of both sexes may be seen close together, frequently flicking open their wings. The males display in flight with rapid wing-fluttering and lift the abdomen tip to reveal a coloured area ('tail light') while hovering close to a female. Eggs are laid into the stems of floating or submerged plants, the females often submerging to do so. Individuals may wander well away from water, especially early in the flight season.

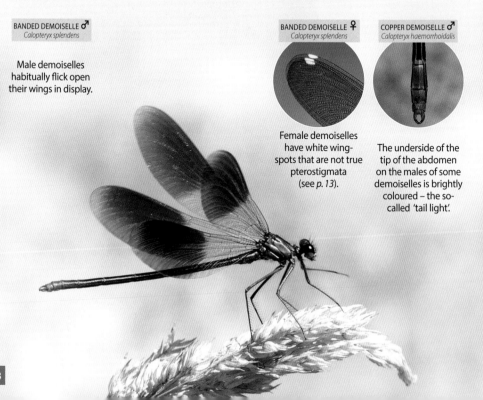

BANDED DEMOISELLE ♂
Calopteryx splendens

Male demoiselles habitually flick open their wings in display.

BANDED DEMOISELLE ♀
Calopteryx splendens

COPPER DEMOISELLE ♂
Calopteryx haemorrhoidalis

Female demoiselles have white wing-spots that are not true pterostigmata (see *p. 13*).

The underside of the tip of the abdomen on the males of some demoiselles is brightly coloured – the so-called 'tail light'.

Demoiselles compared

♂	♀

Beautiful Demoiselle
Calopteryx virgo   (p. 40)

♂ More-or-less all-dark wings (base and tip of wings darker in S of range, darkest in ssp. *festiva*).
♀ Wings brown, some as dark as ♂.

ssp. *virgo*

ssp. *festiva*

Banded Demoiselle
Calopteryx splendens   (p. 42)

♂ Dark band across wings (more extensive in SE Europe).
♀ Wings typically pale green (but some, particularly in SE Europe, have mainly dark brown wings).

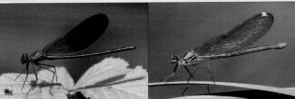

ssp. *splendens*

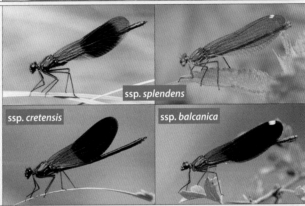

ssp. *cretensis*

ssp. *balcanica*

Western Demoiselle
Calopteryx xanthostoma   (p. 44)

♂ Dark on wings more extensive than on Banded Demoiselle in W Europe. ♀ Usually has brown tip to abdomen.

Copper Demoiselle
Calopteryx haemorrhoidalis
          (p. 46)

Body colour varies from metallic dark reddish-purple to very dark brown, purple or blue.
♂ Wings extensively sooty brown-black.
♀ Dark tip to hindwing.

LC Beautiful Demoiselle

Calopteryx virgo ×2

Beautiful Jewelwing

| Common | = |
| River, stream | |

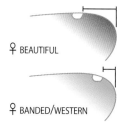

J F M A M J J A S O N D

| Overall length: | 45–49mm |
| Hindwing: | 29–36mm |

♀ BEAUTIFUL

♀ BANDED/WESTERN

COMPARISON OF WING-SPOTS

The largest demoiselle in Europe and also the most widespread, replacing Banded Demoiselle (*p. 42*) where watercourses are fast-flowing. Three subspecies occur.

Identification: Very large. ♂ Body metallic blue-green; underside of abdomen tip dark (brown to reddish). Subspecies differ in wing coloration: *virgo* **dark brown-black** with metallic blue veins and clear base and tip; *meridionalis* as *virgo* but tip dark; and *festiva* **entirely purple-blue**. ♀ Body metallic green with bronzy tip to abdomen; wings translucent **brown** with metallic veins and white wing-spot **farther** from the wingtip than on Banded and Western (*p. 44*) Demoiselles; generally not possible to identify to subspecies but some have darker wings, with similar pattern to male. VARIATION: Extent of male's wing patch varies between subspecies: ssp. *virgo* (N and C Europe) has diffuse pale areas at base and tip of wing, ssp. *meridionalis* (SW) has pale wing-base only and ssp. *festiva* (SE) has wings entirely purple-blue; mixtures of these and intermediates occur in south-central Europe. Immature males have dark brown wings.

Behaviour: Males are territorial, perching on bankside vegetation and trees. They habitually flick open their wings when perched, and dash off to chase passing insects with a distinctive fluttering flight, often returning to same perch. Frequently strays well away from water.

Breeding habitat: Well-oxygenated, flowing waters, often wooded with a sand or gravel base.

LOOK-ALIKES (in range)
Other demoiselles (*pp. 42–47*)

♂ (all subspecies) wings mainly dark with metallic sheen (when mature)

♂ base and tip of wings pale

♂ ssp. *virgo*

♂ ssp. *festiva*

♂ base of wings pale

♂ ssp. *meridionalis*

♂ wings all-dark

♀ ssp. *virgo*

♀ wings brown with white wing-spot away from tip

♀ ssp. *festiva* (dark individual)

LC Banded Demoiselle

Calopteryx splendens ×2

Banded Jewelwing

Common =

River, stream, canal

J F M A M J J A S O N D

Overall length:	45–48mm
Hindwing:	27–35mm

♀ BEAUTIFUL

♀ BANDED/WESTERN

COMPARISON OF WING-SPOTS

A common sight along slow-flowing streams and rivers in the lowlands. Several subspecies have been described based on wing pattern, but due to the considerable variation within populations and across the range these are of limited use.

Identification: Very large. ♂ Body metallic blue-green; underside of abdomen tip ('tail light') yellow to pale grey; wings pale at base, well-defined broad, metallic **dark blue/brownish-black patch** across outer part (extending to wingtip in some subspecies, see *variation*). ♀ Body metallic green; narrow stripe down middle of bronzy-coloured S8–10 usually paler than on Beautiful Demoiselle; wings translucent **pale green** with metallic veins and white wing-spot **nearer** the wingtip than on Beautiful Demoiselle (*p. 40*).
VARIATION: Immatures and some females have similar wing patches to male, but brown; pigmentation in males is evident from emergence. Extent of male's wing patch varies between subspecies: for example, *splendens* (W Europe) has well-defined band and **translucent wingtip**; *balcanica* (SE Europe) has very broad patch extending from node almost to wingtip (to wingtip in some Turkish populations); *caprai* (Italy) has broad patch extending almost to wingtip; and in *cretensis* (Crete) outer half of wing is blue-black to wingtip. Hybrids with Copper Demoiselle (*p. 46*) are known from central Italy.

Behaviour: Males are territorial, but large numbers can be found among lush bankside vegetation and on floating plants. They court females by flicking wings open and performing an aerial dance in front of them.

Breeding habitat: Slow-flowing streams, rivers and canals with muddy sediment. Both sexes frequently wander well away from breeding watercourses and may be found at ponds, where breeding unlikely.

LOOK-ALIKES (in range)
Other demoiselles (*pp. 40–47*)

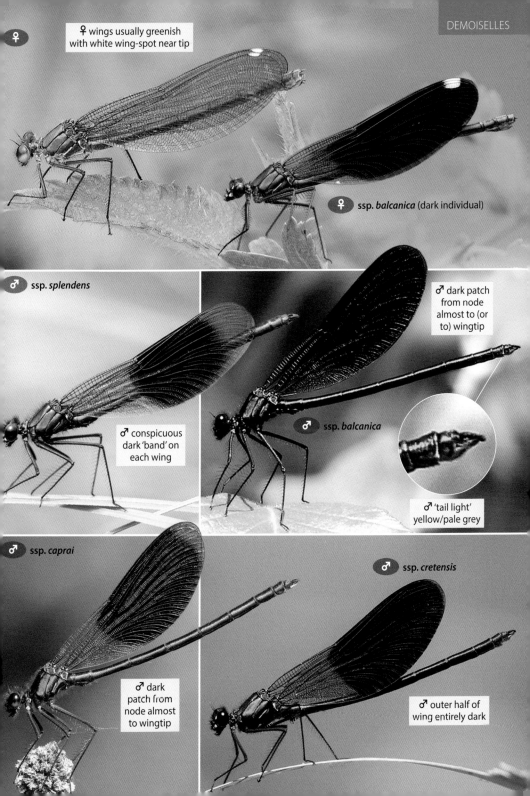

♀

♀ wings usually greenish with white wing-spot near tip

♀ **ssp.** *balcanica* (dark individual)

♂ **ssp.** *splendens*

♂ dark patch from node almost to (or to) wingtip

♂ conspicuous dark 'band' on each wing

♂ **ssp.** *balcanica*

♂ 'tail light' yellow/pale grey

♂ **ssp.** *caprai*

♂ **ssp.** *cretensis*

♂ dark patch from node almost to wingtip

♂ outer half of wing entirely dark

LC Western Demoiselle   *Calopteryx xanthostoma* ×2

Yellow-tailed Demoiselle

Common ═
River, stream, canal

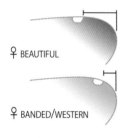

J F M A M J J A S O N D

Overall length:	45–48 mm
Hindwing:	28–31 mm

♀ BEAUTIFUL

♀ BANDED/WESTERN

COMPARISON OF WING-SPOTS

The SW equivalent of Banded Demoiselle (*p. 42*), of which formerly treated as a subspecies, but now separated on the basis of genetic analysis. Range overlaps with Banded Demoiselle in S France and NW Italy, where hybridization occurs.

Identification: Very large; extremely similar to Banded Demoiselle and best identified by location. ♂ Virtually identical to forms of Banded Demoiselle with extensively pigmented wings that occur in SE Europe (but there is no overlap in range). Body metallic blue-green; underside of abdomen tip ('tail light') yellowish; wings pale at base, well-defined broad, metallic **dark blue/brownish-black patch** from node **to tip**. ♀ Body metallic green, but tip of abdomen usually **dull brown**; wings translucent **pale green** with metallic veins and white wing-spot **nearer** the wingtip than in Beautiful Demoiselle (*p. 40*). VARIATION: Immature males develop dark wing pigmentation during first week after emergence (unlike Banded Demoiselle).

Behaviour: Males are territorial and can be found among lush bankside vegetation and on floating plants. Like other demoiselles they display by flicking their wings open.

Breeding habitat: Unshaded to partly shaded, slow- to moderately fast-flowing, well-vegetated streams, rivers and canals in the lowlands.

LOOK-ALIKES (in range)
Other demoiselles (*pp. 40–47*)

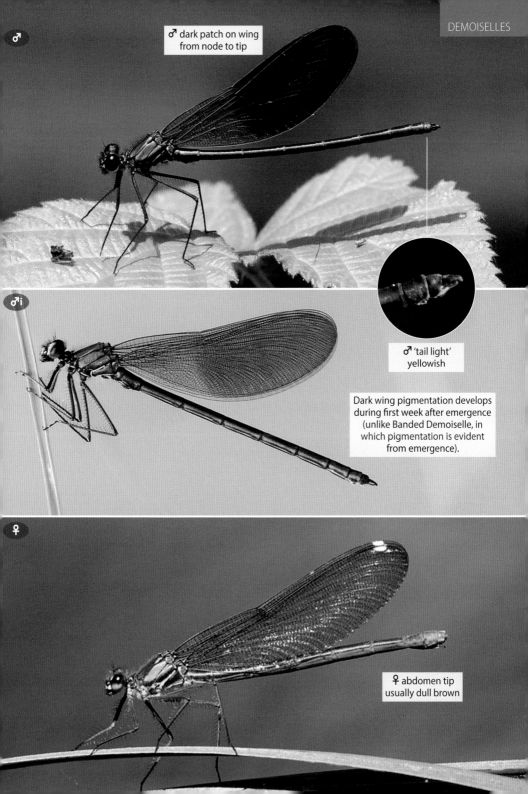

♂ dark patch on wing from node to tip

♂ 'tail light' yellowish

Dark wing pigmentation develops during first week after emergence (unlike Banded Demoiselle, in which pigmentation is evident from emergence).

♀ abdomen tip usually dull brown

LC Copper Demoiselle

Calopteryx haemorrhoidalis ×2

Mediterranean Demoiselle

| Common = |
| Streams, rivers |

J F M A M J J A S O N D

| **Overall length:** | 45–48 mm |
| **Hindwing:** | 23–37 mm |

A western Mediterranean species and arguably the most attractive demoiselle. Like others in the group, it inhabits flowing waters, where males flash open their pigmented wings and occasionally raise the tip of their abdomen in a stunning pink signal to females.

Identification: Very large. Body colour ranges from metallic **dark reddish-purple** to very dark brown, purple or blue. ♂ Wings mostly **sooty brown-black** with small pale patch at base that tapers to a point on rear edge; wings pale at tip (dark on some); underside of abdomen tip ('tail light') bright **pink**. ♀ Thin pale shoulder stripes; wings lightly suffused brown or green, with conspicuous **darker tip to hindwing** (only) and white wing-spot. VARIATION: Extent of dark pigmentation in wing, size, body coloration and behaviour very variable, resulting in ssp. *occasi* (including ssp. *almogravensis*) and ssp. *asturica* being described, but none of these is now considered to be valid. Hybrids with Banded Demoiselle (*p. 42*) are known from central Italy.

Behaviour: Much as other demoiselles, flicking wings open from perches over and beside suitable watercourses and males hovering over water in display to females. Often found in large numbers with other demoiselles. Strays from water and may wander, sometimes well away from breeding areas, even as far as central France.

Breeding habitat: Clear streams and rivers, including smaller and shadier streams than those occupied by other demoiselles.

LOOK-ALIKES (in range)
Other demoiselles (*pp. 40–45*)

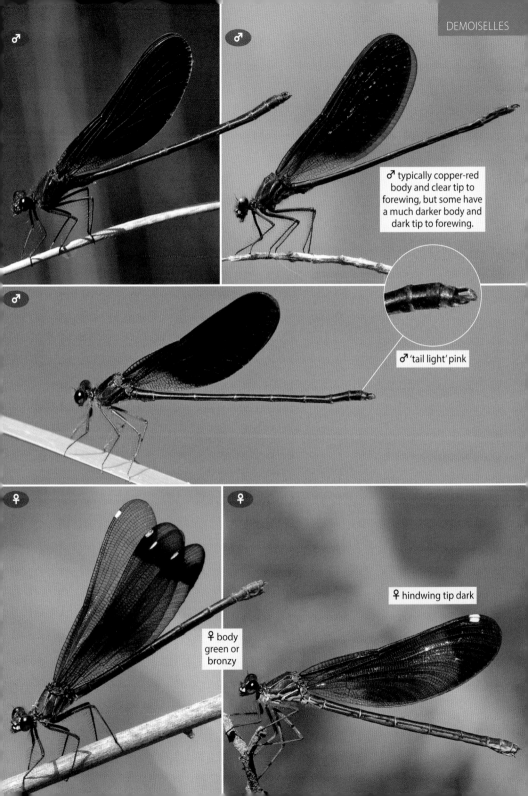

♂ typically copper-red body and clear tip to forewing, but some have a much darker body and dark tip to forewing.

♂ 'tail light' pink

♀ body green or bronzy

♀ hindwing tip dark

ODALISQUE

GENUS | *Epallage*

FAMILY | **Euphaeidae**

1 SPECIES

A large damselfly (male 50 mm; female 40 mm) that, uniquely, holds **wings out** from body like a dragonfly. The only European representative of the Oriental Gossamerwing family. Breeds in flowing waters and lays eggs in tandem.

NT Odalisque

Epallage fatime ×2

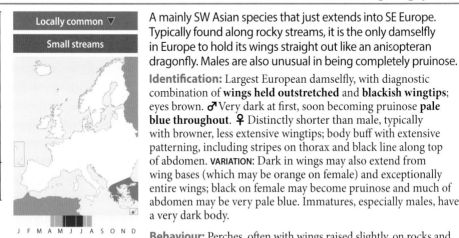

Locally common ▽

Small streams

J F M A M J J A S O N D

Overall length:	♂ 50 mm
	♀ 40 mm
Hindwing:	30–34 mm

A mainly SW Asian species that just extends into SE Europe. Typically found along rocky streams, it is the only damselfly in Europe to hold its wings straight out like an anisopteran dragonfly. Males are also unusual in being completely pruinose.

Identification: Largest European damselfly, with diagnostic combination of **wings held outstretched** and **blackish wingtips**; eyes brown. ♂ Very dark at first, soon becoming pruinose **pale blue throughout**. ♀ Distinctly shorter than male, typically with browner, less extensive wingtips; body buff with extensive patterning, including stripes on thorax and black line along top of abdomen. **VARIATION:** Dark in wings may also extend from wing bases (which may be orange on female) and exceptionally entire wings; black on female may become pruinose and much of abdomen may be very pale blue. Immatures, especially males, have a very dark body.

Behaviour: Perches, often with wings raised slightly, on rocks and the tips of twigs and other vegetation over or close to flowing water. Like demoiselles (*pp. 40–47*), may perch high in trees at times and both sexes can be found in close proximity with little interaction.

Breeding habitat: Fast-flowing waters, especially streams in rocky areas, where tandem pairs lay eggs into stems and woody debris just below waterline. Breeding waters may dry up in summer.

♂

♀

LOOK-ALIKES (in range)
Demoiselles (*pp. 40–47*)
Pruinose skimmers
(*pp. 262–279*)

♂ ♂+♀ Dragonfly-like spread-winged posture and dark wingtips are diagnostic.

♂ completely pruinose when mature

♀

♀i

some females have extensive black in wings

♀

♀

RED DAMSELS

GENERA | *Pyrrhosoma & Ceriagrion*

FAMILY | **Coenagrionidae**

4 SPECIES

Small to medium-sized damselflies (mean lengths 30–38 mm) with at least some **red on abdomen** and one or two black lines across side of thorax. Breed mainly in standing waters; lay eggs in tandem. Two pairs of almost identical 'twins', each pair having one widespread and one highly localized species, the latter occurring in the far SE.

Key features of	GENUS *Pyrrhosoma*	GENUS *Ceriagrion*
◆ EYES	red with **dark horizontal bands**	**red**; no bands
◆ LEGS + WING-SPOTS	**black**	**red** (or reddish)
◆ PALE SHOULDER STRIPES	**red** or **yellow**	**no** obvious stripes
◆ SIDE OF THORAX	**one** black stripe	**two** black lines
◆ ABDOMEN	male red; **bronzy-black** on S7–9	males and some females **all-red**
Identify species by	shape of ♂ appendages | ♀ pronotum	shape of ♂ appendages and fine details of tip of S10 | ♀ pronotum

LC Large Red Damsel

Pyrrhosoma nymphula ×2

Large Red Damselfly, Spring Redtail

Common =

Standing & flowing waters

J F M A M J J A S O N D

Overall length:	33–36 mm
Hindwing:	19–24 mm

A robust and active damselfly, much the more common and widespread of two extremely similar species; the first dragonfly to emerge in spring in the north of its range.

Identification: Medium-sized; thorax bronzy-black above with pale shoulder stripes and black stripe across side; legs and wing-spots **black**; eyes red/brown, crossed by **two dark horizontal stripes**. Almost identical to Greek Red Damsel (*p. 52*), except for marginally smaller size and fine details of male appendages and female pronotum (see *pp. 52–53*). ♂ Abdomen deep red with bronzy-black bands on S7–9; in side view, lower appendages **marginally shorter** than upper, with a 'hook' projecting up **near tip**, about **two-thirds** the length of the upper. ♀ Abdomen ranges from red with limited black markings (forms *typica* and *fulvipes*) to largely black (form *melanotum*); all have narrow yellow bands between segments, becoming **red** towards tip (blue on dark female bluets (*pp. 68–93*) and female redeyes (*pp. 98–101*)); rear edge of pronotum has two **small folds, not raised** in side view. VARIATION: Pale shoulder stripes yellow on immatures.

Behaviour: Males aggressive but not clearly territorial, flying fast in hot conditions. May bask on pale surfaces, including clothing.

Breeding habitat: Wide variety of wetlands, from acidic bog pools to brackish ditches, especially sheltered waters with abundant aquatic plants; avoids fast-flowing waters, but often at flowing waters in E.

LOOK-ALIKES (in range)

Greek Red Damsel (*p. 52*)
Other red damsels (*pp. 54–57*)
Dark ♀ bluets (*pp. 68–93*)
Orange Featherleg (*p. 58*)
♀ redeyes (*pp. 98–101*)

♂

♂ legs and wing-spots black (rules out Small Red Damsel (*p. 54*), the only other widespread red damselfly)

In areas of possible overlap with Greek Red Damsel, check details of male appendages and female pronotum (see *pp. 52–53*).

♂

♀ *typica*

♀ *typica*

♀ *typica*

♀ *fulvipes*

♀ *melanotum*

♀ abdomen ranges from mostly black to mostly red

E CR Greek Red Damsel

Pyrrhosoma elisabethae ×2

Rare and local ▽

Lowland streams

J F M A M J J A S O N D

Overall length:	36–38 mm
Hindwing:	20–24 mm

COMPARISON OF ♀ PRONOTUMS

GREEK RED DAMSEL

two deep folds, raised

LARGE RED DAMSEL

small folds, not raised

Until recently, regarded as a subspecies of the very similar Large Red Damsel (*p. 50*). It is endemic to the S Balkans and one of Europe's rarest species, only known from a few sites in the Peloponnese, Corfu and southern Albania (but is easily overlooked and may occur in as-yet undiscovered locations).

Identification: Medium-sized; thorax bronzy-black above with pale shoulder stripes and black stripe across side; legs and wing-spots **black**; eyes red/brown, crossed by **two dark horizontal stripes**. Almost identical to Large Red Damsel except for marginally larger size and fine details of male appendages and female pronotum. ♂ Abdomen deep red with bronzy-black bands on S7–9; in side view, lower appendages **marginally longer** than upper, with small 'hook' projecting up near base, about **one-third** as long as appendages and **near base** (two-thirds and near tip on Large Red Damsel). ♀ Abdomen either red with limited black markings (form *typica*) or largely black (form *melanotum*); both have narrow yellow bands between segments, becoming **red** towards tip (blue on dark female Azure Bluet (*p. 70*) and Ornate Bluet (*p. 86*)); rear edge of pronotum has two **deep folds, visibly raised** in side view (small folds not raised on Large Red Damsel – see comparison *below*).

Behaviour: Found at water only on sunny days between 11:00 and 16:00, when males take up territorial perches in shafts of sunlight. Unlike Large Red Damsel, males reorient their positions immediately after landing.

Breeding habitat: Not well studied, but occurs along well-vegetated streams in lowland areas (Large Red Damsel is found in mountainous areas in S Balkans). Under severe threat from drought, abstraction and forest fires related to climate change, as well as habitat loss and degradation from direct human action.

LOOK-ALIKES (in range)

Large Red Damsel (*p. 50*)
Dark ♀ Azure Bluet (*p. 70*)
Dark ♀ Ornate Bluet (*p. 86*)

♀ i *melanotum*

♀ i

♀ *melanotum*

Can only be told from Large Red Damsel (*p. 50*) by fine details of ♂ appendages and ♀ pronotum (see comparisons *below* and *opposite*).

♀ *melanotum*

♂

♂ i

COMPARISON OF ♂ APPENDAGES (FROM SIDE)

LARGE RED: lower appendages slightly shorter than upper, with 'hook' near tip

LARGE RED

GREEK RED: lower appendages slightly longer than upper, with small 'hook' near base

GREEK RED

LC Small Red Damsel

Ceriagrion tenellum ×2

Small Red Damselfly, Small Redtail

Common =
Pond, stream, bog, flush

J F M A M J J A S O N D

Overall length:	25–35 mm
Hindwing:	15–21 mm

The commonest and most widespread of two extremely similar very small red damselflies, the other being Turkish Red Damsel (*p. 56*), although the ranges are not known to overlap. It is a weak flyer and rather inconspicuous.

Identification: Very small; both sexes have **red or reddish legs, eyes and wing-spots**; thorax bronzy-black on top, with no more than a hint of pale shoulder stripes, and yellow on side with **two black lines**. Identical to Turkish Red Damsel, except for **smaller** size and fine details of tip of male abdomen and female thorax (see *p. 56*). ♂ Abdomen **entirely bright red.** ♀ Abdomen most commonly, bronzy-black with S1–3 and S9–10 mostly red (form *typica*) but ranges from all-red (form *erythrogastrum*, resembling male) to almost entirely dark, marked with pale segment divisions, last three reddish (form *melanogastrum*) (NOTE: segment divisions blue on female dark form bluets (*pp. 68–93*)); black-marked individuals between *typica* and *erythrogastrum* are known as form *intermedium*.

Behaviour: Flies low and weakly, rarely moving far from breeding waters. Can be surprisingly inconspicuous, especially darker females.

Breeding habitat: Prefers shallow, relatively warm waters, both acidic and calcareous: bog pools, seepages, slow-flowing streams and calcareous mires.

LOOK-ALIKES (in range)
Large Red Damsel (*p. 50*)
Dark ♀ bluets (*pp. 68–93*)
Orange Featherleg (*p. 58*)

♀ melanogastrum

♀ dark form can be confusing – look for pinkish legs, wing-spots and segment divisions at tip of abdomen.

♂

♀ typica

♀i intermedium

♀ erythrogastrum

♂

♂ abdomen, legs, eyes and wing-spots red (rules out Large Red Damsel (*p. 50*), the only other widespread red damselfly)

♂

CR Turkish Red Damsel

Ceriagrion georgifreyi ×2

Rare and local ▽

Streams, seepages

J F M A M J J A S O N D

Overall length:	35–40 mm
Hindwing:	17–20 mm

Rare 'twin' of the almost identical Small Red Damsel (*p. 54*), restricted to a few wetlands in E Mediterranean. Examination with a hand lens is needed to confirm identification.

Identification: Small; both sexes have **red or reddish legs, eyes and wing-spots**; thorax bronzy-black on top, with no more than a hint of pale shoulder stripes, and yellow on side with **two black lines**; identical to Small Red Damsel except for **larger** size and fine details of tip of male abdomen and female thorax. ♂ Abdomen **entirely bright red**; ridge across tip of S10 has **row of tiny, black spines** (hard to see; some Small Red Damsels have a few, but even smaller spines); lower appendages **longer and more slender** than those of Small Red Damsel (*see inset*). ♀ Most commonly, abdomen bronzy-black with S1–3 and S9–10 mostly red (*typica*-like); also has all-red (*erythrogastrum*-like), all-dark (*melanogastrum*-like) forms and possibly others, as Small Red Damsel; at front of thorax, two lobes stick up fairly conspicuously behind slightly raised rear edge of pronotum (absent on Small Red Damsel – see comparison *below*).

Behaviour: As Small Red Damsel, generally keeps rather low, occasionally resting higher up on rushes (*Juncus* spp.) or branches. Seeks shade during the hottest part of the day.

Breeding habitat: Small streams and flushes on islands and mainly coastal areas, even in woodland. Confirmed from Greek islands of Corfu, Thasos and Kakinthos; records of 'Small Red Damsel' from Lesbos and NE Greek mainland could relate to Turkish Red Damsel.

COMPARISON OF TIP OF ♂ ABDOMENS (FROM SIDE)

TURKISH RED DAMSEL

♂ tiny spines on S10
(absent on Small Red Damsel)

SMALL RED DAMSEL

♀ SIDE OF THORAXES FROM SIDE
♀ two lobes at front of thorax behind rear edge of pronotum
(absent on Small Red Damsel)

lobes

no lobes

TURKISH RED DAMSEL   SMALL RED DAMSEL

LOOK-ALIKES (in range)
Large Red Damsel (*p. 50*)

♀ *melanogastrum*-like

♀ *typica*-like

♀ *typica*-like

♀ *typica*-like

♀ can only be told from Small Red Damsel by fine details of the front of the thorax (see comparison *opposite*).

♀ *typica*-like

♂ can only be told from Small Red Damsel by fine details of top of S10 and appendages (see comparison *opposite*).

♂

♂

FEATHERLEGS
GENUS | *Platycnemis*

FAMILY | **Platycnemididae**

3 SPECIES

Medium-sized (mean lengths 35–36 mm); body pale blue, whitish or orange; **double** thin black shoulder stripes (rarely 1); wing-spots rectangular, orange or brown; **tibiae feather-like** (broad and largely white on two species), used in display. Breed mainly in flowing waters; lay eggs in tandem. Two species restricted to W/SW Europe, the other more widespread.

Identify species by body colour | extent of black on tarsi

E **LC** ## Orange Featherleg

Platycnemis acutipennis ×2

Orange White-legged Damselfly

| Locally common = |
| Flowing waters, canals |

J F M A M J J A S O N D

| Overall length: | 34–37 mm |
| Hindwing: | 18–19 mm |

Aptly named W European endemic, found mostly at flowing waters. Males have a diagnostic combination of blue eyes and mainly orange abdomen.

Identification: Medium-sized; thorax creamy with two parallel black shoulder stripes, lower often curved up at front in 'hockey stick' shape; abdomen more-or-less orange; tibiae not noticeably broad (unlike other featherlegs), hind tibiae with incomplete black stripe. ♂ Eyes **blue**; abdomen mostly **orange**, with paired black marks on top of S7–9 only; S10 and appendages whitish; hind tibiae **slightly broadened** (noticeably broad on other featherlegs); wing-spots orange-brown. ♀ Eyes dull blue-brown; abdomen creamy-orange, usually with fine, paired dark markings on all segments; wing-spots pale brown. VARIATION: Lower black shoulder stripe almost absent on some. Immatures have paler abdomen with more restricted dark markings (similar to immature White Featherleg (*p. 60*)).

Behaviour: Shelters in vegetation at or near water.

Breeding habitat: Mainly flowing waters, including large rivers. Sometimes breeds in canals and standing waters.

LOOK-ALIKES (in range)
Other featherlegs (*pp. 60–63*)
Large Red Damsel (*p. 50*)
Small Red Damsel (*p. 54*)

♀ fine paired dark markings on all segments (more obvious than on White Featherleg (p. 60))

♂ hind tibiae slightly broadened, with incomplete black stripe

♀ some individuals are very similar to White Featherleg (p. 60)

♂ unmistakable combination of blue eyes and orange abdomen

♀ hind tibiae with incomplete black stripe

♂ + ♀ lower black shoulder stripe can be almost absent

Ⓔ LC White Featherleg

Platycnemis latipes ×2

Common	=

Flowing waters

J F M A M J J A S O N D

Overall length:	33–37 mm
Hindwing:	18–22 mm

Endemic to SW Europe and often abundant in marginal vegetation along large rivers, looking strikingly pale.

Identification: Medium-sized; thorax and abdomen generally paler than other featherlegs; thorax with two pairs of parallel black shoulder stripes, lower often curved up at front in 'hockey stick' shape; mid- and hind tibiae **with little or no black**; wing-spots brownish, shorter than those of Blue Featherleg (*p. 62*). ♂ Eyes pale blue; thorax very pale blue-green; abdomen whitish with **no black** on S2–5 apart from (on some) a pair of fine flecks towards end of each segment; mid- and hind tibiae **very broad**; wing-spots orange-brown. ♀ Cream-coloured; lower dark shoulder stripe often discontinuous or absent; paired dark markings along top of abdomen usually faint; wing-spots pale brown. VARIATION: Extent of dark markings on abdomen variable; some females have a buffish head, thorax, base of legs and abdomen tip.

Behaviour: Tends to fly later in summer than other featherlegs.

Breeding habitat: Typically large, lowland rivers with slow to medium flows, but sometimes streams and canals.

COMPARISON OF LEGS

White Featherleg ♂ + ♀ mid- and hind tibiae with little or no black (**Blue Featherleg** usually has black stripe); mid- and hind tibiae conspicuously broad in ♂ of both species.

WHITE FEATHERLEG

BLUE FEATHERLEG

LOOK-ALIKES (in range)

Other featherlegs (*pp. 58–63*)

♀ lower black shoulder stripe often discontinuous or absent

♂ no black along top of S2–5

♀ hind tibiae with little or no black

♂ + ♀ largely white body may lead to confusion with recently emerged Blue Featherleg (p. 62) or Orange Featherleg (p. 58)

♀ some individuals are very similar to Orange Featherleg (p. 58)

♂ mid- and hind tibiae very broad, with little or no black (see opposite)

♂ dark variant

♂ + ♀ wing-spots brownish

[LC] Blue Featherleg

Platycnemis pennipes ×2

White-legged Damselfly

Common =

Flowing/standing waters

J F M A M J J A S O N D

Overall length:	35–37 mm
Hindwing:	19–23 mm

The most widespread featherleg in Europe, breeding mainly in slow-flowing waters, often with Banded Demoiselle (*p. 42*).

Identification: Medium-sized; thorax with two pairs of parallel black shoulder stripes, lower usually curved up at front in 'hockey stick' shape; abdomen with **paired** black markings on **S1–9**; mid- and hind tibiae usually **with black stripe**; wing-spots pale brownish (longer than those of White Featherleg (*p. 60*)). ♂ Body **pale blue**; S7–10 mostly black above; mid- and hind tibiae **very broad**; wing-spots orange-brown. ♀ Body very pale yellow-green or blue-green; wing-spots pale brown. VARIATION: Thorax greenish on some males. Immatures creamy-white with brownish thorax and eyes, and much reduced black abdominal markings, especially on females.

Behaviour: Often congregates in tall waterside vegetation, although immatures may wander far from breeding sites. Males hover and flash their white legs in aggressive display. Sometimes found in large numbers laying eggs in tandem into emergent stems, floating leaves and occasionally rotten wood.

Breeding habitat: Found mostly along muddy rivers and streams in N Europe, favouring unshaded, unpolluted, slow-flowing sections with abundant emergent and floating vegetation. Also occurs at canals, lakes and ponds where margins are oxygenated by wave action, especially in the S.

COMPARISON OF LEGS

White Featherleg ♂ + ♀ mid- and hind tibiae with little or no black (**Blue Featherleg** usually has black stripe); mid- and hind tibiae conspicuously broad in ♂ of both species.

WHITE FEATHERLEG

BLUE FEATHERLEG

LOOK-ALIKES (in range)

Other featherlegs (*pp. 58–61*)
Bluets (*pp. 68–93*)

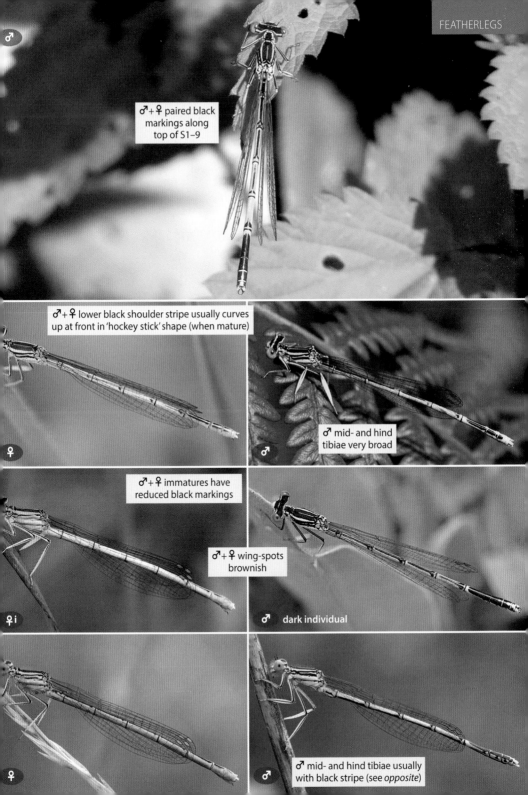

♂

♂ + ♀ paired black markings along top of S1–9

♀

♂ + ♀ lower black shoulder stripe usually curves up at front in 'hockey stick' shape (when mature)

♂ mid- and hind tibiae very broad

♂

♀ i

♂ + ♀ immatures have reduced black markings

♂ + ♀ wing-spots brownish

♂ dark individual

♀

♂ mid- and hind tibiae usually with black stripe (see *opposite*)

♂

BLUETS 1/2

GENERA | *Enallagma & Coenagrion*

FAMILY | **Coenagrionidae**

13 SPECIES

Small to medium-sized (mean lengths 29–36 mm), blue damselflies with complex black patterns (males only), pale shoulder stripes and small, usually black wing-spots.

This section provides an introduction to the potentially confusing species of small damselfly that have predominantly blue males. Two of these species are common and widespread across most of Europe, nine are more localized and the other three are rare. Males of these species (and the similar Blue-eye *Erythromma lindenii*) are shown for comparison opposite, annotated to indicate the critical identification features, together with their respective markings on the second abdominal segment (S2). Black markings on males may vary across the species' ranges, especially in isolated populations. Males are generally more easily found than females, which can be very difficult to identify in isolation. Females are covered on the next two pages and in the individual species accounts that follow.

Key features of bluets

◆ small-medium, delicate damselflies
◆ blue-and-black (mature males)
◆ not metallic or pruinose
◆ relatively weak flight
◆ wing-spots short and plain

Identify species by patterning on head, thorax and abdomen | colour and shape of wing-spots | shape of pronotum | shape of eye-spots

Key identification features

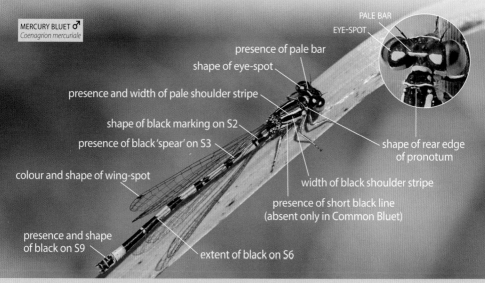

MERCURY BLUET ♂
Coenagrion mercuriale

PALE BAR

EYE-SPOT

presence of pale bar

shape of eye-spot

presence and width of pale shoulder stripe

shape of black marking on S2

presence of black 'spear' on S3

colour and shape of wing-spot

shape of rear edge of pronotum

width of black shoulder stripe

presence of short black line (absent only in Common Bluet)

presence and shape of black on S9

extent of black on S6

Other types of damselfly with blue markings

LEGS: hind tibiae broad

Blue Featherleg ♂
Platycnemis pennipes (p. 62)

LARGE REDEYE ♂
Erythromma najas

S9–10: ± blue

EYES: red or bright blue

Brighteyes
Erythromma (p. 96)

COMMON BLUETAIL ♂
Ischnura elegans

WING-SPOT: typically two-tone

Bluetails
Ischnura (p. 102)

Male bluets compared (plus Blue-eye)

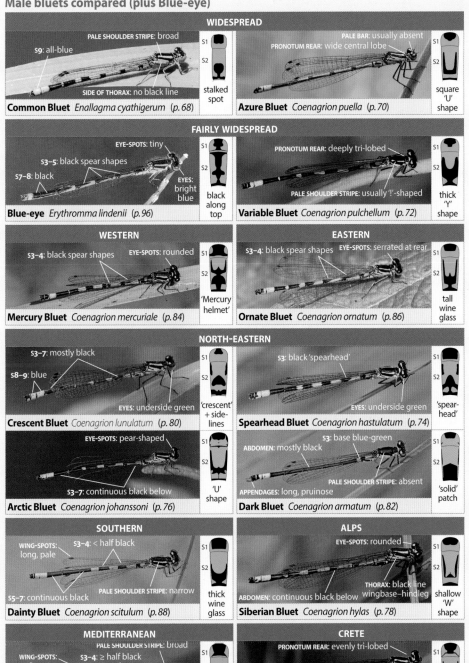

WIDESPREAD

PALE SHOULDER STRIPE: broad
S9: all-blue
SIDE OF THORAX: no black line
S1
S2
stalked spot
Common Bluet *Enallagma cyathigerum* (p. 68)

PALE BAR: usually absent
PRONOTUM REAR: wide central lobe
S1
S2
square 'U' shape
Azure Bluet *Coenagrion puella* (p. 70)

FAIRLY WIDESPREAD

EYE-SPOTS: tiny
S3–5: black spear shapes
S7–8: black
EYES: bright blue
S1
S2
black along top
Blue-eye *Erythromma lindenii* (p. 96)

PRONOTUM REAR: deeply tri-lobed
PALE SHOULDER STRIPE: usually '!'-shaped
S1
S2
thick 'Y' shape
Variable Bluet *Coenagrion pulchellum* (p. 72)

WESTERN

S3–4: black spear shapes
EYE-SPOTS: rounded
S1
S2
'Mercury helmet'
Mercury Bluet *Coenagrion mercuriale* (p. 84)

EASTERN

S3–4: black spear shapes **EYE-SPOTS:** serrated at rear
S1
S2
tall wine glass
Ornate Bluet *Coenagrion ornatum* (p. 86)

NORTH-EASTERN

S3–7: mostly black
S8–9: blue
EYES: underside green
S1
S2
'crescent' + side-lines
Crescent Bluet *Coenagrion lunulatum* (p. 80)

S3: black 'spearhead'
EYES: underside green
S1
S2
'spear-head'
Spearhead Bluet *Coenagrion hastulatum* (p. 74)

EYE-SPOTS: pear-shaped
S1
S2
'U' shape
S3–7: continuous black below
Arctic Bluet *Coenagrion johanssoni* (p. 76)

S3: base blue-green
ABDOMEN: mostly black
PALE SHOULDER STRIPE: absent
APPENDAGES: long, pruinose
S1
S2
'solid' patch
Dark Bluet *Coenagrion armatum* (p. 82)

SOUTHERN

WING-SPOTS: long, pale
S3–4: < half black
S5–7: continuous black **PALE SHOULDER STRIPE:** narrow
S1
S2
thick wine glass
Dainty Bluet *Coenagrion scitulum* (p. 88)

ALPS

EYE-SPOTS: rounded
THORAX: black line wingbase–hindleg
ABDOMEN: continuous black below
S1
S2
shallow 'W' shape
Siberian Bluet *Coenagrion hylas* (p. 78)

MEDITERRANEAN

PALE SHOULDER STRIPE: broad
WING-SPOTS: almost triangular
S3–4: ≥ half black
S5–7: continuous black
S1
S2
thick wine glass
Mediterranean Bluet *C. caerulescens* (p. 90)

CRETE

PRONOTUM REAR: evenly tri-lobed
S1
S2
thick 'U' shape
Cretan Bluet *Coenagrion intermedium* (p. 92)

BLUETS 2/2

GENERA | *Enallagma & Coenagrion*

FEMALES   Female bluets often lack blue coloration and the distinctive abdominal patterning shown by males, which are usually present nearby (and often in tandem). Females may occur in dark forms (where much of the abdomen is black), as well as blue, green or brown forms. Blue forms (known as androchrome) suggest males, but of course females possess an ovipositor, which bulges below the tip of the abdomen. Details of the rear edge of the pronotum may be needed to confirm identification.

Identify species by checking for males nearby or ideally in tandem | shape of rear edge of pronotum | patterning on thorax, abdomen and head | colour and shape of wing-spots

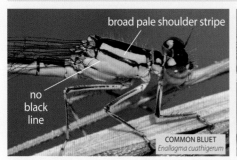

broad pale shoulder stripe

no black line

COMMON BLUET
Enallagma cuathigerum

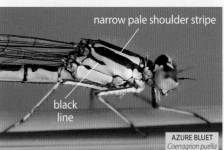

narrow pale shoulder stripe

black line

AZURE BLUET
Coenagrion puella

Common Bluet (*left*) differs from all *Coenagrion* bluets by the absence of a short black line on the side of the thorax, in addition to the pale shoulder stripe being broader than the black line below.

Identification of female bluets

The identification of female bluets is best confirmed by the shape of the rear edge of the pronotum. This can be seen on captured individuals using a hand lens (or binoculars used in reverse!) but is sometimes possible from high quality photos.

Common Bluet
Enallagma cyathigerum
(*p. 68*)

Azure Bluet *Coenagrion puella* (*p. 70*)	**Variable Bluet** *Coenagrion pulchellum* (*p. 72*)	**Mercury Bluet** *Coenagrion mercuriale* (*p. 84*)	**Ornate Bluet** *Coenagrion ornatum* (*p. 86*)
Spearhead Bluet *Coenagrion hastulatum* (*p. 74*)	**Crescent Bluet** *Coenagrion lunulatum* (*p. 80*)	**Siberian Bluet** *Coenagrion hylas* (*p. 78*)	**Arctic Bluet** *Coenagrion johanssoni* (*p. 76*)
Dainty Bluet *Coenagrion scitulum* (*p. 88*)	**Mediterranean Bluet** *Coenagrion caerulescens* (*p. 90*)	**Dark Bluet** *Coenagrion armatum* (*p. 82*)	**Cretan Bluet** *Coenagrion intermedium* (*p. 92*)

Female bluets compared (plus Blue-eye)

WIDESPREAD

BROWN FORM **s3–7:** black 'rocket' shapes
s8: black triangle

PALE BAR: usually absent
BLUE FORM
<30% blue
PRONOTUM REAR: weakly tri-lobed

SIDE OF THORAX: no black line   **PALE SHOULDER STRIPE:** broad

Common Bluet *Enallagma cyathigerum* (p. 68)

Azure Bluet *Coenagrion puella* (p. 70)

FAIRLY WIDESPREAD

s3–7: blunt 'rocket' shapes
APPENDAGES: pale

PALE BAR: present
BLUE FORM
<30% blue

s3–6: sides blue   **PALE SHOULDER STRIPE:** broad

Blue-eye *Erythromma lindenii* (p. 96)

Variable Bluet *Coenagrion pulchellum* (p. 72)

WESTERN | EASTERN

DARK FORM   **PRONOTUM:** +/- straight; small central protrusion

EYE-SPOTS: serrated at rear
BLUE FORM
s4–8: black line through blue bases

s3: black 'rocket' shape

Mercury Bluet *Coenagrion mercuriale* (p. 84)

Ornate Bluet *Coenagrion ornatum* (p. 86)

NORTH-EASTERN

s4–8: blue patch at base

PRONOTUM REAR: blunt point

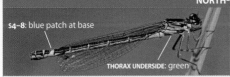

THORAX UNDERSIDE: green

Crescent Bluet *Coenagrion lunulatum* (p. 80)

Spearhead Bluet *Coenagrion hastulatum* (p. 74)

BLUE FORM
s10: blue

s8: blue-green base   **s2:** blue-green base
ABDOMEN: mostly black

s3–7: continuous black below

Arctic Bluet *Coenagrion johanssoni* (p. 76)

Dark Bluet *Coenagrion armatum* (p. 82)

SOUTHERN | ALPS

BLUE FORM   **WINGSPOTS:** long, pale

s8: two blue patches at base   **EYE-SPOTS:** rounded

THORAX: black line
wingbase–hindleg
s8: blue patch at base   **s3–4:** black 'rocket' shapes
ABDOMEN: continuous black below   **THORAX UNDERSIDE:** green

Dainty Bluet *Coenagrion scitulum* (p. 88)

Siberian Bluet *Coenagrion hylas* (p. 78)

MEDITERRANEAN | CRETE

BLUE FORM   **WINGSPOTS:** almost triangular
s3–7: black 'rocket' shapes
s10: blue

PALE BAR: present
s3–8: blue bases

PALE SHOULDER STRIPE: broad

PRONOTUM REAR: evenly tri-lobed

Mediterranean Bluet *C. caerulescens* (p. 90)

Cretan Bluet *Coenagrion intermedium* (p. 92)

LC Common Bluet

Enallagma cyathigerum ×2

Common Blue Damselfly

| Common | = |

| Lake, pond, river |

J F M A M J J A S O N D

| **Overall length:** | 29–36 mm |
| **Hindwing:** | 18–20 mm |

S2   S1

♂ S2: stalked spot

♀ pronotum: shallow curve

The most widely distributed and often most abundant damselfly in Europe, found in a wide range of standing or slow-flowing waters, where males characteristically patrol low and well out over the water.

Identification: Both sexes easily told from other bluets by **lack** of obvious short black line on side of thorax (present on all *Coenagrion* bluets – see *p. 66*) and **broad** pale shoulder stripes (narrow on all other bluets except rare Mediterranean Bluet (*p. 90*)). ♂ Abdomen usually has black, **stalked spot on S2** and completely **blue S8–9** (apart from two small black spots on S9); blue fades to grey at low temperature. ♀ Occurs in dull green, brown and blue colour forms; shapes of black marks on abdomen include 'thistle' on S2, various **'rocket'** shapes on S3–7 and **triangle** on S8. Unlike other blue damselflies, has obvious **spine** in front of the ovipositor, under S8. VARIATION: Shape of S2 variable on males, as is extent of black generally. Pale areas are straw-coloured on immatures.

Behaviour: Eggs laid into vegetation at and below surface, male releasing female when she submerges. Males aggressive in pursuit of females and patrol over water looking for females emerging from water after egg-laying, which are 'rescued' and helped back to shore. In poor weather, large numbers shelter in tall marginal vegetation. Immatures often disperse far from water.

Breeding habitat: Very wide range of open waters, both standing and slow-flowing, only really avoiding small ponds and ditches, and fast-flowing waters, but can occur anywhere. Tolerates both nutrient-poor, acidic lakes (with no fishes) and enriched conditions, with huge populations at many large lakes and reservoirs.

LOOK-ALIKES (in range)
Other bluets (*pp. 70–93*)
Blue-eye (*p. 96*)
Blue Featherleg (*p. 62*)
Winter damsels (*pp. 34–37*)

♀ vulvar spine under S8: (lacking in other bluets)

♂+♀ lack of short black line on side of thorax diagnostic

♀

♂i

♂+♀ pale shoulder stripes broad

♀

♂ blue S8–9, side to thorax and shoulder stripes make it look bluer than other bluets

♂ S9 all-blue

♀ blue form

♀ S3–7 'rocket'-shaped markings on top

♂

LC Azure Bluet

Coenagrion puella ×2

Azure Damselfly

Common =

Pond, canal, ditch

J F M A M J J A S O N D

Overall length:	33–35 mm
Hindwing:	16–23 mm

S2   S1

♂ S2: square 'U' shape

♀ pronotum: wide central lobe

One of the two commonest bluets in Europe (the other being Common Bluet (*p. 68*)), except in the far north and south of its range. It is especially at home in small ponds and ditches, and is the benchmark against which to compare less common bluets.

Identification: At close range, both sexes (and all other *Coenagrion* bluets) can be distinguished from Common Bluet by presence of short black line on side of thorax (see *p. 66*) and, with the exception of rare Mediterranean Bluet (*p. 90*), pale shoulder stripes being **narrower** than black line below. **Lacks** pale bar between eye-spots (present in most other bluets). As with other bluets, identification of both sexes can be confirmed by shape of rear edge of pronotum. ♂ Blue abdomen has isolated black '**U**' **shape** on S2; black markings extend forwards on sides of S3–6; paired or joined black markings on rear half of S9. ♀ Generally blue-green with extensive black on abdomen. **VARIATION:** Some variation in black markings on S2 and S9 on males; female also occurs in blue form that has blue extending to <30% on S4–5; immatures have blue areas whitish.

Behaviour: Often seen in large numbers egg-laying in tandem into vegetation at or just below water's surface. Rarely ventures far out over large areas of open water (unlike Common Bluet).

Breeding habitat: Wide range of standing waters, including acidic and eutrophic, but favours smaller, more sheltered sites. Regularly found in garden ponds and small ditches (where Common Bluet usually absent).

LOOK-ALIKES (in range)
Other bluets (*pp. 68–93*)
Blue-eye (*p. 96*)
Blue Featherleg (*p. 62*)

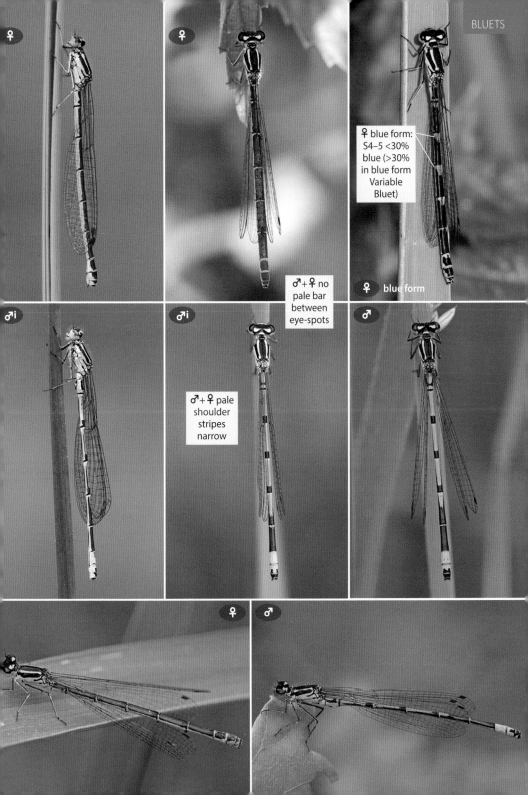

♀

♀

♀ blue form:
S4–5 <30%
blue (>30%
in blue form
Variable
Bluet)

♀ blue form

♂+♀ no
pale bar
between
eye-spots

♂i

♂i

♂

♂+♀ pale
shoulder
stripes
narrow

♀

♂

LC **Variable Bluet**

Coenagrion pulchellum ×2

Variable Damselfly

| Common | = |

| Lake, pond, canal, ditch, fen, bog | |

J F M A M J J A S O N D

| Overall length: | 33–38 mm |
| Hindwing: | 16–23 mm |

S2　　S1

♂ S2: thick 'Y' shape

♀ pronotum: deeply tri-lobed

Widespread except in the far N and SE, but with a patchy distribution and generally rarer than the similar Azure Bluet (*p. 70*). Despite its name, it is not the only bluet prone to variations in patterning.

Identification: Both sexes have pale bar between eye-spots and distinctive **three-lobed** rear edge to pronotum, with prominent narrow middle lobe (especially on female). ♂ Pale shoulder stripes typically **form an 'exclamation mark' ('!') shape**; abdomen has thick black **'Y' shape** (like wine goblet in cross-section) on S2. ♀ Abdomen typically dark like other female bluets. VARIATION: Males highly variable across range, some paler like Azure Bluet, others with more extensive black (*e.g.* in S Europe and Turkey); pale shoulder stripes may be complete or absent; S2 marking may lack 'stem'. Female has blue form with blue extending to >30% on S4–5 (<30% on Azure Bluet); locally, dark females may show largely blue S8 (resembling Dark Bluet (*p. 82*)).

Behaviour: Similar to Azure Bluet. Adults are often found in lush vegetation a short distance from water.

Breeding habitat: Well-vegetated often alkaline standing waters, such as grazing marsh ditches, fens, ponds, lakes and canals. Sometimes found in bogs and rarely in slow-flowing waters.

LOOK-ALIKES (in range)
Other bluets (*pp. 68–93*)
Blue-eye (*p. 96*)
Blue Featherleg (*p. 62*)

♀ blue form: S4–5 >30% blue (<30% in blue form Azure Bluet)

♀ blue form

♂

♂ dark variant

♀

♀ best identified by shape of hind margin of pronotum (see *opposite*)

♂

♀

♀ pale bar between eyes is useful distinction from Azure Bluet (*p. 70*)

♂ pale shoulder stripes usually '!'- shaped

♀

♀i

♀ some otherwise dark individuals have mostly blue S8

♀ blue form

LC Spearhead Bluet

Coenagrion hastulatum ×2

Northern Damselfly

Common **?**

Lake, pond, bog

J F M A M J J A S O N D

Overall length:	31–33 mm
Hindwing:	17–22 mm

S2   S1

♂ S2: 'spearhead' shape with detached side-lines (variable)

♀ pronotum: blunt central point

One of the commonest bluets in NE Europe, becoming more localized and occurring in mountains farther S and W, as far as Bulgaria and the Pyrenees.

Identification: Both sexes have a pale bar between eye-spots; eyes and 'face' have **bright green undersides** and pronotum has **blunt point** in middle of rear edge. ♂ Abdomen has S8–9 blue with variable black markings on S9 and, on some, S8; S2 has **black 'spearhead'** shape, usually linked to division between S2 and S3 and detached from black line along each side; S3 has black **'spearhead'** shape; lower appendages longer than upper, sharply incurved towards tip. ♀ Top of abdomen mostly black above, black widening gradually towards tip. VARIATION: Male S2 'spearhead' shape variable in extent and not always linked to S2–3 division.

Behaviour: Flies rather weakly, rarely venturing out over open water; tends to favour sedge beds, usually being found low down close to water's surface.

Breeding habitat: Shallow (<60 cm deep) pools and sheltered margins of acidic lakes and ponds with dense beds of emergent sedges (*Carex* spp.) and horsetails (*Equisetum* spp.).

LOOK-ALIKES (in range)
Other bluets (*pp. 68–93*)
Blue-eye (*p. 96*)
Blue Featherleg (*p. 62*)

♂

♀ + ♂ bright green undersides to eyes and 'face'

♀

♂

♀

♂ + ♀ pale bar between eye-spots

♂

♂ S2 typically has black 'spearhead' and detached side-lines, but the shape is variable (see *top right*)

♂ S3 black 'spearhead'

♂ extent of black markings on abdomen variable

♀i

♂i

LC Arctic Bluet

Coenagrion johanssoni ×2

| Local | = |
| Lake, bog | |

J F M A M J J A S O N D

| Overall length: | 27–30 mm |
| Hindwing: | 15–19 mm |

S2   S1

♂ S2: 'U' shape (variable)

♀ pronotum: shallow curve

A small, dark bluet that has the most northerly range of any European damselfly. It has extensive black markings on the body, which probably enables more rapid warming in cool conditions.

Identification: Small bluet with **continuous black** marking along lower sides of **S3–7**. This feature shared only with much larger and very rare Siberian Bluet (*p. 78*), but differs in having pear-shaped (not rounded) eye-spots with pale bar between. Overlaps widely in range with Spearhead Bluet (*p. 74*) and Crescent Bluet (*p. 80*) but lacks green underside. Rear edge of pronotum almost straight. ♂ Black markings on top of abdomen similar to Azure Bluet (*p. 70*), but 'U' shape on S2 thicker and, viewed from side, corners extend in a point back to bottom of segment division. ♀ Blue with extensive black markings; **S10 and tip of S9 blue. VARIATION:** Some males have S2 marking 'broken' at corners (as Spearhead Bluet) or with line back to rear edge of segment (as Variable Bluet (*p. 72*)); S8–9 may have pairs of faint or dark spots; some males have black line on side of thorax from base of hindwing to base of hind leg; females may have blue areas brown.

Behaviour: Much as other bluets. Adults keep close to vegetation.

Breeding habitat: Marshy fringes of ponds, lakes and moorland bogs in the boreal zone (taiga and rarely in tundra).

LOOK-ALIKES (in range)
Other bluets (*pp. 68–93*)
Blue Featherleg (*p. 62*)

♀ brown form

Damselflies often clean debris such as spiders' web from their wings by raising and lowering their abdomen.

♂ ♂ + ♀ eye-spots pear-shaped

♀ blue form

♀ S10 and tip of S9 blue

♂

♀ blue form

♂ ♂ continuous black markings along lower side of S3–7

VU Siberian Bluet

Coenagrion hylas ×2

Frey's Damselfly, Bilek's Damselfly

PROTECTED

Very rare =

Lake, pond

J F M A M J J A S O N D

Overall length:	33–38 mm
Hindwing:	19–28 mm

S2   S1

♂ S2: shallow 'W' shape

♀ pronotum: straight with small central protrusion

The largest European bluet, found in the Lech River valley, a very restricted area of the northern Tyrol in Austria, where it is isolated from the species' main Siberian range. Has more extensive black markings on the body than most bluets, probably an adaptation to warming up more quickly in cool conditions.

Identification: Large bluet with continuous black markings along **entire side** of abdomen; eye-spots **rounded**, with pale bar between; **black line** on side of thorax from base of hindwing to base of hind leg (like some Arctic Bluets (*p. 76*)). Rear edge of pronotum straight with small central protrusion. ♂ S2 marking very shallow 'W'; black marks on S3–6 are broader than on Azure Bluet, with fine, forward-pointing **spikes**; S7 has a **divided blue** base. ♀ Underside of thorax and abdomen tip **pale green**; abdomen as Dainty Bluet, with black **'rocket' shapes** on top of S3–6, pairs of blue spots at fronts of S7–8, and S9–10 mostly black. **VARIATION:** Base of male's S2 marking may be triangular and detached from sides; S8–9 may have limited dark markings.

Behaviour: Not surprisingly for an Alpine species, most active in fine weather around middle of the day.

Breeding habitat: Small, shallow lakes and ponds in northern Alpine valleys at 800–1,600 m. These are clear, cold, oligotrophic, densely fringed with sedges and fed by water from calcareous springs.

LOOK-ALIKES (in range)
Other bluets (*pp. 68–93*)

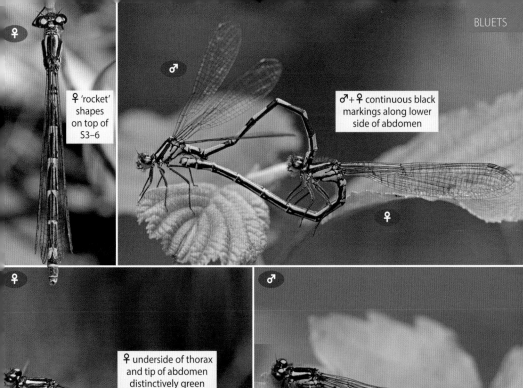

♀ 'rocket' shapes on top of S3–6

♂ + ♀ continuous black markings along lower side of abdomen

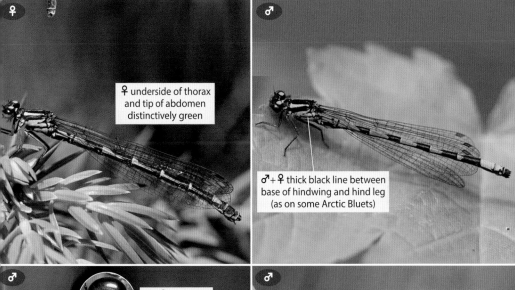

♀ underside of thorax and tip of abdomen distinctively green

♂ + ♀ thick black line between base of hindwing and hind leg (as on some Arctic Bluets)

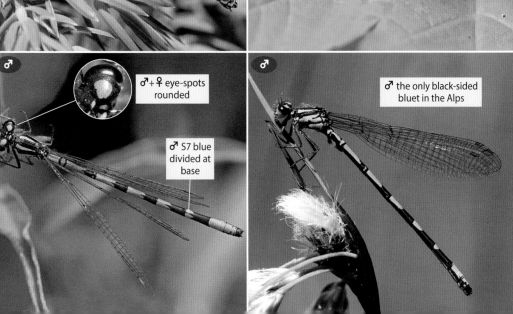

♂ + ♀ eye-spots rounded

♂ S7 blue divided at base

♂ the only black-sided bluet in the Alps

LC Crescent Bluet

Coenagrion lunulatum ×2

Irish Damselfly, Irish Bluet

Local and uncommon ?

Lake, pond

J F M A M J J A S O N D

Overall length:	30–33 mm
Hindwing:	16–22 mm

S2   S1

♂ S2: 'crescent' plus side-lines

♀ pronotum: clear central protrusion

A rather localized bluet of NE Europe, with a few outposts farther S and in Ireland.

Identification: Both sexes have green eyes and **raised central lobe** protruding on rear edge of pronotum. ♂**Lacks** pale bar between eye-spots; abdomen rather dark blue, usually **apple-green** below; S2 marking usually isolated black **'crescent'** with detached black line along each side; S3–7 **mostly black**; **S8–9 blue**. ♀ Green/brown; narrow pale bar between eye-spots; abdomen largely black, like dark forms of other bluets, but front end of S8 **blue, interrupted by black line**. VARIATION: Male S8–9 may have pairs of small black dots and black bars at tips.

Behaviour: Adults shelter in tall vegetation close to water, males often flying out to sit on floating leaves. Pairing takes place away from water, females found by water almost invariably being accompanied by males.

Breeding habitat: Shallow, nutrient-poor acidic ponds and lakes, often in or near woodland, with clear water, submerged and floating plants and open beds of fringing emergent plants, such as sedges (*Carex* spp.). In E of range, also in more eutrophic ponds, gravel pits, *etc.*

LOOK-ALIKES (in range)

Other bluets (*pp. 68–93*)
Blue-eye (*p. 96*)
Blue Featherleg (*p. 62*)

♀

♂

♂ S3–7 mostly black;
S8–9 usually all-blue

♀

♂

♂ no pale bar between eye-spots

♀ front of
S8 blue,
interrupted
by black line

♀ i

♂

♂

♂ + ♀ underside
of eyes green

NT Dark Bluet

Coenagrion armatum ×2

Norfolk Damselfly

Scarce and local ▽

Lake, pond, river, ditch

J F M A M J J A S O N D

| Overall length: | 31–34 mm |
| Hindwing: | 16–20 mm |

S2   S1

♂ S2: 'solid' black patch plus side-lines

♀ pronotum: green sides; prominent pale central lobe

An unusually dark bluet, which at first glance may resemble a bluetail (*pp. 102–119*) or redeye (*pp. 98–101*) rather than a bluet. Occurs locally from the Netherlands eastwards, but has been lost from many sites, including Great Britain. Main range from the Baltic eastwards. Believed lost from the Netherlands in 20th century, but rediscovered in 1999.

Identification: Slender, delicate species; both sexes have distinctive **black** abdomen with **coloured** areas at both **base and tip**. ♂ **Lacks** pale shoulder stripes **and** pale bar between eye-spots; abdomen blue-green on S1–3 and S8–9 only; S2 has relatively large black patch abutting S3, with detached side-lines; S8–9 have limited black markings; **exceptionally long**, incurved, spatulate lower appendages, **pruinose** and about twice length of S10. ♀ Pale bar between eye-spots; pale shoulder stripes green; pronotum green at sides with prominent **pale lobe** in the middle of rear edge; abdomen **black** with **blue-green** on S1 and basal halves of S2 and S8. VARIATION: A few males have spots where the pale shoulder stripe would be; S2 side-lines joined to 'solid' black patch in some; immatures pinkish.

Behaviour: Inconspicuous, keeping low. Males fly low but strongly through relatively open beds of emergent vegetation; may perch on floating leaves.

Breeding habitat: Favours shallow ponds and lakes with beds of low-density Common Reed (*Phragmites australis*), sedges (*Carex* spp.) and Water Horsetail (*Equisetum fluviatile*), tolerating only limited amount of nutrient enrichment; sometimes ditches and sluggish rivers.

LOOK-ALIKES (in range)
Dark ♀ Variable Bluet (*p. 72*)
Bluetails (*pp. 102–119*)
Redeyes (*pp. 98–101*)

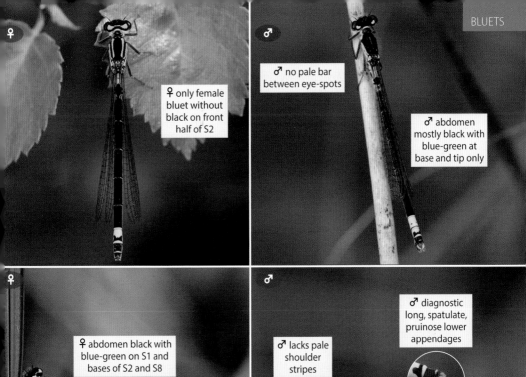

♀ only female bluet without black on front half of S2

♂ no pale bar between eye-spots

♂ abdomen mostly black with blue-green at base and tip only

♀ abdomen black with blue-green on S1 and bases of S2 and S8

♂ lacks pale shoulder stripes

♂ diagnostic long, spatulate, pruinose lower appendages

♀ pale bar between eye-spots

NT Mercury Bluet

Coenagrion mercuriale ×2

Southern Damselfly

PROTECTED

Locally common in SW, rare in N & E of range ▽

Stream, ditch, flush

J F M A M J J A S O N D

Overall length:	29–31 mm
Hindwing:	15–20 mm

S2   S1

♂ S2: 'Mercury helmet' mark

♀ pronotum: small central lobe

The smallest bluet, occurring on flowing waters in S and W Europe; the western counterpart of Ornate Bluet (*p. 86*).

Identification: Both sexes have a pale bar between rather rounded eye-spots and small lobe in centre of rear edge of pronotum. ♂ Best identified by **'Mercury helmet' mark on S2**; black **spear-shaped markings** on S3–4; S9 more than half black; upper appendages as long as lower appendages, and incurved. ♀ Usually dull green with abdomen mainly black above, with blue divisions between last few segments (similar dark form of Small Red Damsel (*p. 54*) has red segment divisions). **VARIATION:** Male S2 pattern can vary, extremes having 'prongs' narrow and detached. Female can occur in a blue form.

Behaviour: Sedentary species, rarely moving more than 50 m when mature, although some disperse up to 2 km. Flight weak and low. Females usually lay eggs alone into soft-stemmed submerged herbs.

Breeding habitat: Shallow, unshaded gravelly streams, runnels and flushes, often calcareous (or at least slightly base-enriched), with shallow peat or silt over gravel and slow to moderate flows; typically vegetated with soft-tissue plants (for egg-laying) and in grazed grassland, fenland or heathland. Others occur on chalk streams and ditches, and in poor fens. Permanent flows and stable water temperatures are important, as is nearby shelter for adults.

LOOK-ALIKES (in range)

Other bluets (*pp. 68–93*)
Dark ♀ Small Red Damsel (*p. 54*)
Blue-eye (*p. 96*)
Blue Featherleg (*p. 62*)

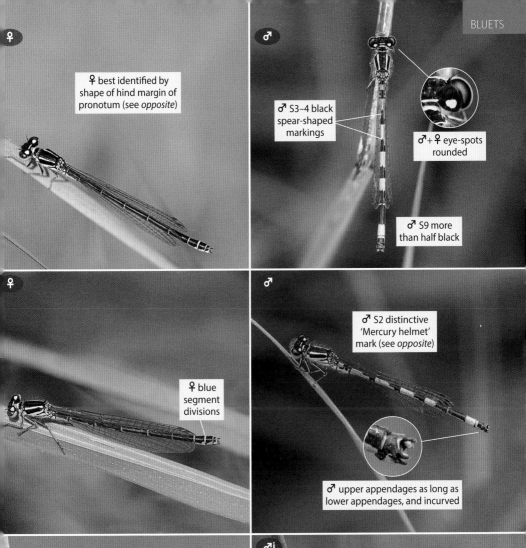

♀ ♀ best identified by shape of hind margin of pronotum (see *opposite*)

♂ ♂ S3–4 black spear-shaped markings

♂+♀ eye-spots rounded

♂ S9 more than half black

♀ ♀ blue segment divisions

♂ ♂ S2 distinctive 'Mercury helmet' mark (see *opposite*)

♂ upper appendages as long as lower appendages, and incurred

♂

♀ blue form

♂i

NT Ornate Bluet

Coenagrion ornatum ×2

PROTECTED

Uncommon ▽

Stream

J F M A M J J A S O N D

| Overall length: | 30–31 mm |
| Hindwing: | 17–24 mm |

S2　　S1

♂ S2: tall 'wine glass' shape
(in cross-section)

♀ pronotum: indented, with
tiny notch in central lobe; large
pale spot on each side

A small, uncommon bluet, very much the eastern counterpart of Mercury Bluet (*p. 84*), using similar habitats and overlapping in range only in limited parts of France and Germany. Identification ideally needs confirmation of serrated eye-spots together with detail of abdominal patterning.

Identification: Small size, short, diamond-shaped wing-spots and usually distinctive **serrated** rear edge to eye-spots; usually a complete pale bar between the eye-spots. ♂ Abdomen has stalked 'wine glass' mark on S2 and pointed markings on S3–7, those on S3–4 **spear-shaped**; S8 blue, usually with two tiny black spots. ♀ Typically blue at bases of S4–8 divided by black line and S3 with pointed **'rocket' shape**; pronotum has indented rear edge with tiny notch in central lobe and pale spots on sides that are larger than on most bluets. VARIATION: Extent of black on abdomen variable on males; dark form females similar to other dark form bluets, but rare.

Behaviour: Flies weakly and low through wetland vegetation.

Breeding habitat: Small, slow-flowing, shallow streams and ditches, typically sunny and calcareous with diverse plant species and structure.

LOOK-ALIKES (in range)
Other bluets (*pp. 68–93*)
Blue-eye (*p. 96*)
Blue Featherleg (*p. 62*)

♀

♀ S4–8 typically have black line dividing blue base

♂

♂ S3–4 black spear-shaped markings

♂ + ♀ rear edge to eye-spots usually serrated

♀

♂

♀

♂

♀ S3 'rocket'-shaped marking

♀

Coenagrion scitulum ×2

Dainty Damselfly

Fairly common =

Pond, ditch

J F M A M J J A S O N D

Overall length:	33–38 mm
Hindwing:	16–23 mm

S2   S1

♂ S2: thick 'wine glass' shape
(in cross-section)

♀ pronotum: tri-lobed; small
blue spot on each outer lobe

A delicate bluet similar to the very local Mediterranean Bluet (*p. 90*), but unlike that species is rarely found at flowing waters.

Identification: Small bluet with pale bar between eye-spots, **narrow** pale shoulder stripes and, on some, an isolated black dot at tip of short black line on side of thorax. Both sexes have **pale** wing-spots that are **longer** than on other bluets (nearly twice as long as wide). ♂ On abdomen, S2 marking typically resembles stalked wine glass, with sides diverging; S5–7 effectively 2½ continuous black segments (most other bluets have less than 2); S8–9 are blue (some with a little black on S9); and S10 mostly black; upper appendages longer than lower, but both **shorter** than those of Mercury Bluet (*p. 84*). ♀ Typically bright blue with black markings; S3–4 with black '**rocket**' shapes, like Common Bluet (*p. 68*), but **lacks** broad pale shoulder stripes and vulvar spine of that species; rear edge of pronotum **tri-lobed**, like Variable Bluet (*p. 72*), but with **blue spot** on each outer lobe. **VARIATION:** Black pattern on males variable; females may be duller with yellowish shoulder stripes.

Behaviour: Males often sit on floating vegetation well away from the shore, investigating any approaching damselflies. Egg-laying takes place in tandem into surface pondweeds or debris.

Breeding habitat: Open standing waters with beds of pondweed, typically water-milfoils (*Myriophyllum* spp.) and hornworts (*Ceratophyllum* spp.). It can tolerate temporary (in S Europe) and slow-flowing waters, and moderate levels of salinity.

Comparison of wing-spots

pale, long (± 2× as long as wide)   pale, ± triangular, tapers towards wingtip

DAINTY BLUET   MEDITERRANEAN BLUET

LOOK-ALIKES (in range)
Other bluets (*pp. 68–93*)
Blue-eye (*p. 96*)
Blue Featherleg (*p. 62*)

♂ S3–4 < half black

♂ S5–7 2½ continuous black markings

♂ S8–9 blue

♂ + ♀ pale shoulder stripes narrow

♀ S3–4 'rocket'-shaped markings

♂ + ♀ wing-spots relatively long and pale (see *opposite*)

NT Mediterranean Bluet

Coenagrion caerulescens ×2

J F M A M J J A S O N D

Overall length:	30–33 mm
Hindwing:	14–21 mm

S2   S1

♂ S2: thick 'wine glass' shape (variable)

♀ pronotum: two projecting spurs, with deep cleft

This small, west Mediterranean species is rare in southern France, but commoner farther south, where it is only to be found at flowing waters. Males and females are very similar to those of the much commoner Dainty Bluet (*p. 88*).

Identification: Both sexes have typically **large**, pear-shaped blue eye-spots with a blue bar between; pronotum has **large**, blue spot on each side (small on Dainty Bluet) and the rear edge has two projecting **spurs bisected by deep cleft**; blue shoulder stripes **as wide as or broader** than the black stripe below; wing-spot is pale, **almost triangular** in shape and tapers acutely towards wingtip. ♂Abdomen has S2 mark like thick wine glass, with thickened base emphasised by bulges and relatively short side extensions; S3–4 at least half black (less than half on Dainty Bluet); S5–7 effectively 2½ continuous black segments (most other bluets have less than 2); typically less black on S9–10 than other bluets. ♀ S3–7 with 'rocket' shapes; S9 with black lines separated by blue in middle; S10 **blue**. VARIATION: some populations (*e.g.* in Italy) have more extensive areas of black, including tip of abdomen on females.

Behaviour: Flies low and weakly amongst lush emergent plants (as Mercury Bluet (*p. 84*)).

Breeding habitat: Well-vegetated shallow running waters, both natural and artificial, in hot, sunny locations.

Comparison of wing-spots

pale, long (± 2× as long as wide)   pale, ± triangular, tapers towards wingtip

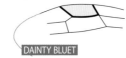

DAINTY BLUET   MEDITERRANEAN BLUET

LOOK-ALIKES (in range)
Other bluets (*pp. 68–93*)
Blue-eye (*p. 96*)
Blue Featherleg (*p. 62*)

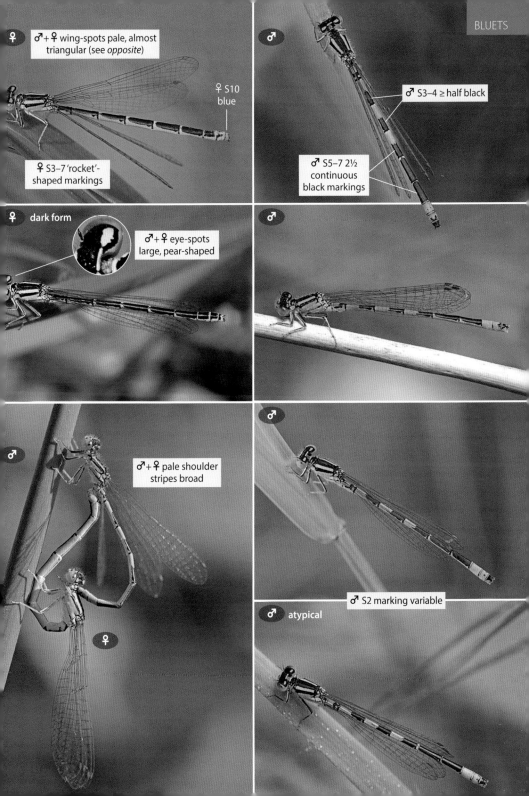

♀ ♂ + ♀ wing-spots pale, almost triangular (see *opposite*)

♀ S10 blue

♀ S3–7 'rocket'-shaped markings

♂ S3–4 ≥ half black

♂ S5–7 2½ continuous black markings

♀ dark form

♂ + ♀ eye-spots large, pear-shaped

♂

♂

♂ + ♀ pale shoulder stripes broad

♂

♀

♂ S2 marking variable

♂ atypical

E **VU** **Cretan Bluet** *Coenagrion intermedium* ×2

Rare and local ?

Stream

J F M A M J J A S O N D

| Overall length: | 35–36 mm |
| Hindwing: | 19–22 mm |

S2   S1

♂ S2: thick 'U' shape (variable)

♀ pronotum: evenly tri-lobed
(central lobe more prominent
than on Azure Bluet)

As its name suggests, this species is endemic to Crete, where it is the most widely distributed, if localized, bluet. Unlike other bluets, it appears to breed solely in running water, perhaps a result of the scarcity of permanent standing waters in Crete. It closely resembles Azure Bluet (*p. 70*) in appearance.

Identification: Both sexes similar to Azure Bluet, but pale bar between eye spots and rear edge of pronotum evenly tri-lobed (intermediate between Azure and Variable Bluets (*p. 72*)). ♂ Blue abdomen has isolated black 'U' shape on S2 thicker than on Azure Bluet; black markings extend forwards on sides of S3–6; black on S9 can cover more, or less, than half of segment (less than half on Azure Bluet, except on the Greek mainland where black more extensive). ♀ Abdomen mainly black with black thistle shape on S2 (like Azure Bluet); blue bases to S3–8, broader than on typical dark form female Azure Bluet. VARIATION: On some males the long, thin sides of the S2 marking are separated from the black base; black on S8 less than half of segment on some.

Behaviour: Tandem pairs lay eggs into streamside vegetation, including roots, algae and dead leaves.

Breeding habitat: Unusually for a bluet, inhabits permanent flowing waters. This may be in response to the natural scarcity of standing fresh water on this limestone-dominated island. It is vulnerable to climate change and abstraction from permanent watercourses.

LOOK-ALIKES (in range)
Common Bluet (*p. 68*)
Blue-eye (*p. 96*)

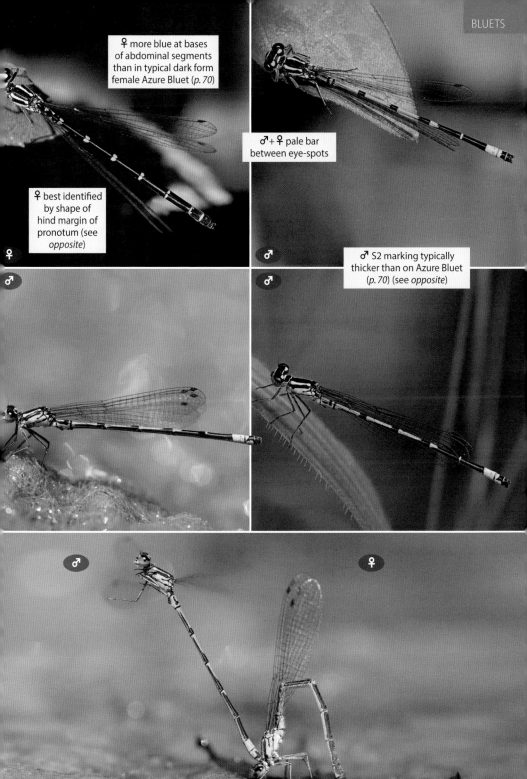

♀ more blue at bases of abdominal segments than in typical dark form female Azure Bluet (*p. 70*)

♀ best identified by shape of hind margin of pronotum (see *opposite*)

♂ + ♀ pale bar between eye-spots

♂ S2 marking typically thicker than on Azure Bluet (*p. 70*) (see *opposite*)

'BLUE-TAILED' DAMSELFLIES

FAMILY | **Coenagrionidae**

GENERA | *Erythromma, Ischnura & Nehalennia*   13 SPECIES

Small to medium-sized (mean lengths 24–33 mm) damselflies, the males of which have a 'blue tip' (actually somewhere on S8–10) to a mostly dark abdomen.

This section provides an introduction to a group of small to medium-sized damselflies, the males of which have blue patches towards the tip of their abdomen. Five of these species are common and widespread across much of Europe, while the remainder are restricted in range and sometimes highly localized. Males are shown for comparison opposite, annotated to indicate the critical identification features. Males are generally more easily found than females, which can be very difficult to identify in isolation. Females mostly have a dark abdomen and some have a blue 'tail', as the males.

Brighteyes (*Erythromma*) (*pp. 96–101*)

Males of the three brighteyes comprise two species with red eyes, dark body and blue tip to the abdomen, while the third has blue eyes and is blue with black markings. All three have blue 'tail lights' more-or-less covering the tip of the abdomen on S9–10. The redeyes can be differentiated by the extent of blue on the abdomen, Small Redeye having a blue 'wedge' on the side of S8 that is lacking on Large Redeye. Blue-eye resembles a bluet (see *pp. 68–93*), but has blue right at the tip of the abdomen and, like the redeyes, often sits on short perches or floating vegetation well away from the shoreline.

Key features of brighteyes

◆ males have blue S9–10
◆ males patrol/perch over open water

Identify species by eye colour | extent of blue on abdomen

Bluetails (*Ischnura*) (*pp. 102–119*)

Males of the nine bluetails are very similar, with mostly black upperparts and blue 'tail lights' around S8 (*i.e.* not right at the tip). Only Common and Small Bluetails are widespread, the others being very localized. The ranges of several similar species overlap and reliable identification can only be made by studying the pronotum and male appendages of several individuals in the hand (the latter are described in the species accounts). All but one (Citrine Forktail) have a two-toned wing-spot on the forewing, the part nearest to the wingtip being paler than the rest; the difference in tone is less marked on females. Wing-spots are at least twice as long as wide and all the same size, except in two species, in which they are less than twice as long as wide and the hindwing spots smaller. Females occur in a confusing array of colour forms, some of which are immature; some, but not all, have 'tail lights' and/or clear pale shoulder stripes. In the species accounts, each of these forms is referred to simply by the colour of the thorax. Females have a vulvar scale under S8; unlike other damselflies they always lay eggs alone.

Key features of bluetails

◆ abdomen of males and many females mostly dark with blue near tip
◆ colour of female thorax variable, in part with age
◆ wing-spots on forewings two-toned (except Citrine Forktail)

Identify species by extent of blue around S8 | pronotum shape

Sedgling (*Nehalennia speciosa*) (*p. 120*)

Europe's smallest damselfly, a very localized species of north-east Europe found in acidic bogs and lake fringes with beds of tall sedges (*Carex* spp.) and similar dense plants.

Key features of Sedgling

◆ small size
◆ short wings
◆ metallic green above
◆ pale tip to abdomen

Male 'blue-tailed' damselflies compared (Sedgling, brighteyes & bluetails)

With one exception, males of all European 'blue-tailed' damselflies have blue at or near the tip of the abdomen. Only Citrine Forktail *Ischnura hastata* (*p. 118*) 'breaks the rule' but since only females have been recorded in Europe, confusion is unlikely.

For each of the bluetail species (*below*), illustrations of the wing-spot and pronotum are shown.

BRIGHTEYES

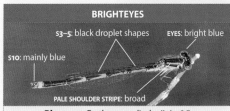

S3–5: black droplet shapes   EYES: bright blue
S10: mainly blue
PALE SHOULDER STRIPE: broad

Blue-eye *Erythromma lindenii* (*p. 96*)

THORAX: top all-dark
S10: all blue

Large Redeye *Erythromma najas* (*p. 98*)

S10: mainly blue   PALE SHOULDER STRIPE: obscure
ABDOMEN: narrow
S8: side blue

Small Redeye *Erythromma viridulum* (*p. 100*)

SEDGLING

UPPERPARTS: metallic green or bronzy   SIZE: tiny
S8–10: top blue

Sedgling *Nehalennia speciosa* (*p. 120*)

BLUETAILS

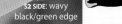

RANGE: widespread
PRONOTUM: 'spike' on rear edge

Common Bluetail *Ischnura elegans* (*p. 106*)

RANGE: Iberia
PRONOTUM: no 'spike' on rear edge

Iberian Bluetail *Ischnura graellsii* (*p. 108*)

RANGE: Tuscan archipelago, Tyrrhenian and Maltese islands
FACE, THORAX AND BASE OF ABDOMEN: green
PRONOTUM: short, weakly divided lobe on rear edge

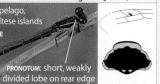

Island Bluetail *Ischnura genei* (*p. 110*)

RANGE: Pantelleria, Italy
S1–2: side blue
THORAX: side blue
S3–7: side orange

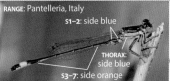

Oasis Bluetail *Ischnura fountaineae* (*p. 116*)

RANGE: Canary Islands
S2 SIDE: straight black/green edge
S3–7: side orange

Sahara Bluetail *Ischnura saharensis* (*p. 112*)

RANGE: Canary Islands (Tenerife, La Palma)
S3–7: side orange   S2 SIDE: wavy black/green edge

Marsh Bluetail *Ischnura senegalensis* (*p. 114*)

RANGE: widespread
WING-SPOT: relatively short
TIP S8+S9: blue with variable black markings

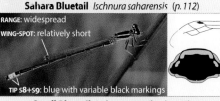

Small Bluetail *Ischnura pumilio* (*p. 102*)

RANGE: Cyprus
WING-SPOT: relatively short
S8+S9: mostly blue

Persian Bluetail *Ischnura intermedia* (*p. 104*)

LC Blue-eye

Erythromma lindenii ×2

Goblet-marked Damselfly

Common △

Lake, river, stream, canal

J F M A M J J A S O N D

Overall length:	34–39mm
Hindwing:	18–23mm

Males typically perch on short stems just above the water's surface.

Easily overlooked due to its similarity to Common Bluet (*p. 68*). Although formerly regarded as a bluet, it is now placed in the same genus as the redeyes (and hence referred to as a 'brighteye'). Occurs from the Mediterranean northwards, but is scarcer N of the Alps, although it has spread northwards in recent years.

Identification: Both sexes have pale shoulder stripes **broader** than the black stripe below, like Common Bluet; eye-spots small, often slit-like, with bar in between; thorax with short black line on side (as *Coenagrion* species); wing-spots **long, pointed** and **yellow-brown**. ♂ Eyes bright blue; black abdominal markings resemble 'droplets' of viscous liquid dripping down from S1; S7–8 **wholly black**, S9–10 **mainly blue** and S10 split by a black line; upper appendages **long** and **pincer-like**. ♀ Abdomen yellow, brownish or green, with **blue** sides to central segments; dark markings on top of S3–7 similar to Common Bluet's 'rocket' shapes, but lacking point at front; S9–10 often pale; appendages **pale**. VARIATION: Eye-spots absent in some; short black line on side of thorax may have a detached dot at end; immatures devoid of blue.

Behaviour: As with Common Bluet, males fly low over open water, perching horizontally just above the surface on short stems.

Breeding habitat: Well-oxygenated slow-flowing rivers, stream, canals, lakes and gravel pits with abundant pondweed.

LOOK-ALIKES (in range)
Bluets (*pp. 68–93*)
Blue Featherleg (*p. 62*)

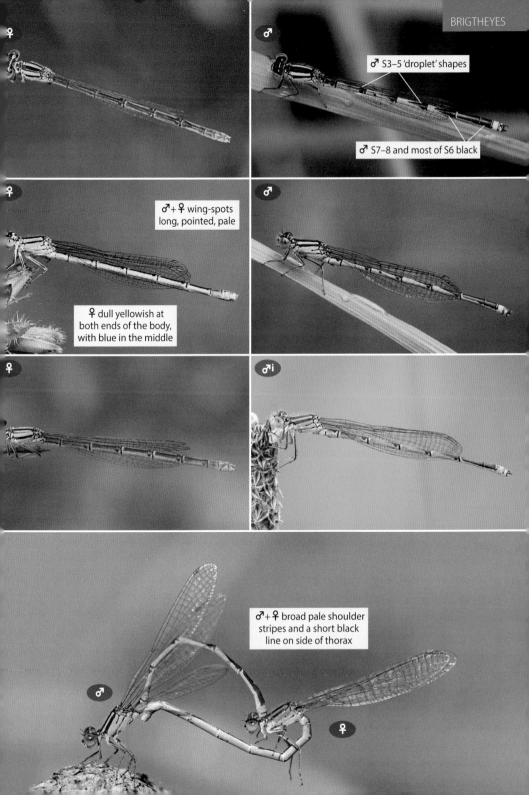

♀

♂

♂ S3–5 'droplet' shapes

♂ S7–8 and most of S6 black

♀

♂ + ♀ wing-spots
long, pointed, pale

♂

♀ dull yellowish at
both ends of the body,
with blue in the middle

♀

♂i

♂ + ♀ broad pale shoulder
stripes and a short black
line on side of thorax

♂

♀

LC Large Redeye

Erythromma najas ×2

Red-eyed Damselfly

| Common = |
| Lake, pond, river, canal |

J F M A M J J A S O N D

| Overall length: | 30–36 mm |
| Hindwing: | 19–24 mm |

COMPARISON OF ♂ REDEYE
ABDOMEN TIPS

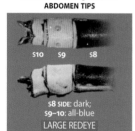

S10   S9   S8

S8 SIDE: dark;
S9–10: all-blue
LARGE REDEYE

S10   S9   S8

S8 SIDE: blue 'wedge';
S9 blue; **S10:** black 'X'
SMALL REDEYE

A robust, dark, 'blue-tailed damselfly', usually found sitting away from the shore on floating vegetation, particularly water-lilies. Widespread over much of Europe except the far S and N, with a more northerly distribution than similar Small Redeye (*p. 100*).

Identification: Larger and bulkier than Small Redeye and bluetails (*pp. 102–119*); both sexes lack pale eye-spots and have pale brown wing-spots, black legs and two black lines on side of thorax, the upper one short, as on *Coenagrion* species (see *p. 66*); wings extend more than halfway along S7. ♂ Eyes deep **burgundy-red**; thorax with **top completely bronzy-black**, **lacking pale shoulder stripes**, and side blue; abdomen has blue S1 and **S9–10** and dark side to S8, otherwise dark on top with slight pruinescence. ♀ Eyes brownish-red; **narrow, short or discontinuous** yellowish shoulder stripes; otherwise yellowish-green below and dark above with blue divisions to last few abdominal segments; rear edge of pronotum strongly **tri-lobed**. VARIATION: Immatures like female but bronzy with coloured areas yellowish; some males have a detached black spot at end of the short black line on side of thorax.

Behaviour: Flies earlier in the season than Small Redeye, peaking about a month earlier, and usually keeps abdomen straight. In fine weather, males patrol low over water or sit on floating leaves, where they fight for strategic positions near open areas. They quickly move to nearby vegetation when the sun goes in, often landing in trees. Eggs are laid, while in tandem, into stems and leaves of floating and sometimes emergent plants. Egg-laying often underwater, still in tandem.

Breeding habitat: Closely associated with floating leaves, typically water-lilies, but also pondweeds and other floating vegetation. Favoured sites include larger ponds, lakes and flooded mineral workings, canals, large drains and slow-flowing rivers.

LOOK-ALIKES (in range)

Small Redeye (*p. 100*)
Bluets (*pp. 68–93*)

♀ pale shoulder stripes incomplete

♂ top of thorax uniformly bronzy-black

♂ eyes burgundy-red (colour not always easy to see)

♂ S9–10 all-blue, side of S8 dark (see *opposite*)

♀ top of S10 dark

♂i

♀i

♂ + ♀ wings extend > halfway along S7

♂

LC Small Redeye

Erythromma viridulum ×2

Small Red-eyed Damselfly

Common △

Lake, pond, ditch, canal

J F M A M J J A S O N D

Overall length:	26–32 mm
Hindwing:	16–20 mm

COMPARISON OF ♂ REDEYE ABDOMEN TIPS

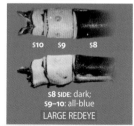

S10 S9 S8

S8 SIDE: dark;
S9–10: all-blue
LARGE REDEYE

S10 S9 S8

S8 SIDE: blue 'wedge';
S9 blue; **S10:** black 'X'
SMALL REDEYE

A small version of Large Redeye (*p. 98*) and also likely to be seen on floating vegetation. Although it has a more southerly distribution than that species, its range has expanded rapidly to the N recently. Numbers peak a month later than Large Redeye, coinciding with the growth of pondweed at the surface.

Identification: Like Large Redeye, both sexes lack pale spots behind the eyes and have pale brown wing-spots and two black lines on side of thorax, the upper one short, as on *Coenagrion* bluets (see *p. 66*), which on males often ends in a dot (rarely on Large Redeye); wings reach no more than halfway along S7.
♂ Eyes **brownish-red**; thorax bronzy-black on top with blue side; pale shoulder stripes complete or reduced; abdomen dark except for blue on the top of S1, S9–10 and the **sides of S2–3 and S8**; **black 'X'-shape on top of S10**; tip of abdomen often upcurved.
♀ Eyes brown above, greenish below; shoulder stripes **complete** and, like side of thorax, are yellow, green or blue; abdomen similar to that of Large Redeye, except that top of S10 is **pale** (blue, green or yellow), extending forward just onto top and side of S9; rear edge of pronotum is **rounded**. VARIATION: Females may become lightly pruinose.

Behaviour: Mating occurs either on floating plants or at the margins. When perched on floating vegetation, males hold their abdomen slightly upcurved (usually straight in Large Redeye). Eggs are laid, while in tandem, into floating plants.

Breeding habitat: Frequents a range of standing and sometimes slow-flowing waters with floating mats of pondweed, particularly hornworts (*Ceratophyllum* spp.) or water-milfoils (*Myriophyllum* spp.); sometimes on lily-pads with Large Redeye. Tolerates brackish conditions.

LOOK-ALIKES (in range)
Large Redeye (*p. 98*)
Bluets (*pp. 68–93*)

♀ some individuals may become pruinose

♂ S9 and 'wedge' on side of S8 blue; black 'X' on top of S10 (see *opposite*)

♀ pale shoulder stripes complete or reduced

♂ eyes brownish-red (colour not always easy to see)

♀ top of S10 pale

♂ top of thorax bronzy-black with complete or reduced pale shoulder stripes

♂ + ♀ wings extend ≤ halfway along S7

♂ abdomen often curves upwards at the tip

LC Small Bluetail

Ischnura pumilio ×2

Scarce Blue-tailed Damselfly, Small Bluetip

| Locally common = |
| Pond, stream, ditch |

NORTH
SOUTH
J F M A M J J A S O N D

| **Overall length:** | 26–31 mm |
| **Hindwing:** | 14–18 mm |

♂ pronotum: arched, without distinct central lobe

COMPARISON OF WINGSPOTS

| < 2× as long as wide | ≥ 2× as long as wide |
| SMALL BLUETAIL | COMMON BLUETAIL |

Small, unobtrusive and easily overlooked, this species favours sparsely vegetated, warm, shallow, often ephemeral waters. As a pioneer species of new sites, it is highly mobile and populations are often short-lived. There has been some northward expansion in range in response to climate change.

Identification: Forewings of both sexes have **two-toned**, diamond-shaped wing-spots, **shorter** than those of Common Bluetail (*p. 106*), < 2× as long as wide, and noticeably larger than those on hindwings; rear edge of pronotum arched, lacking a distinct central lobe. ♂ Blue or blue-green eyes, side of thorax and narrow shoulder stripes; abdomen mainly black on top apart from blue on **S9** and tip of S8. ♀ Greenish eyes and side of thorax; **broad brownish-green** shoulder stripes with very narrow black stripe below; abdomen black above, lacking pale S8; wing-spots browner and less clearly two-toned than male. VARIATION: Male S9 often has pair of small, but variable, black marks; females rarely in blue form; coloured parts paler and greener on immature males; immature females extensively **orange**, extending to wing veins and top of S1–3.

Behaviour: Despite weak appearance, disperses over long distances by using favourable air-currents when plant density reaches a critical stage. Has at least two generations per year. Eggs laid mainly into emergent grasses (Poaceae) and rushes (*Juncus* spp.) above or just below waterline.

Breeding habitat: Occurs in small streams, flushes and ponds, often associated with mineral extraction sites. Waters are shallow and usually sparsely vegetated. Ephemeral sites, such as rainwater pools, settlement lagoons and even wheel ruts, may be suitable, but usually only for a limited time.

LOOK-ALIKES (in range)
Other bluetails (*pp. 104–119*)
Redeyes (*pp. 98–101*)

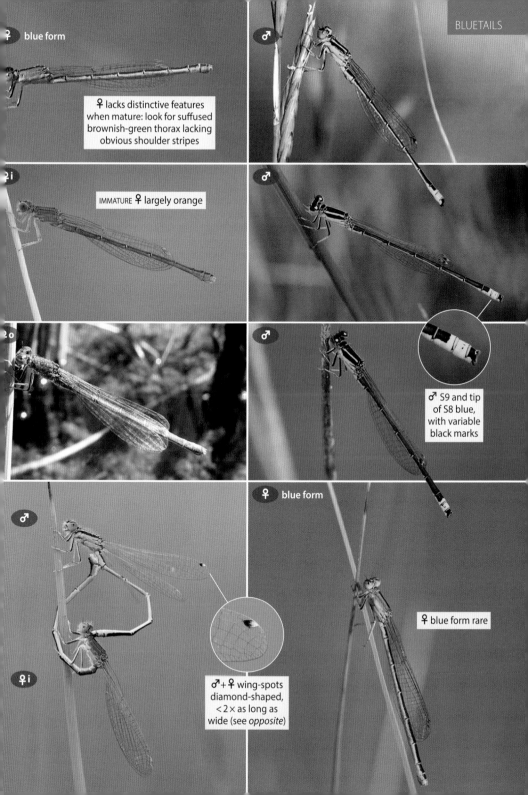

♀ blue form

♂

♀ lacks distinctive features when mature: look for suffused brownish-green thorax lacking obvious shoulder stripes

2 i

IMMATURE ♀ largely orange

♂

2 o

♂

♂ S9 and tip of S8 blue, with variable black marks

♂

♀ blue form

♀ blue form rare

♀ i

♂ + ♀ wing-spots diamond-shaped, <2× as long as wide (see *opposite*)

NE Persian Bluetail

Ischnura intermedia ×2

Dumont's Bluetail

| Rare and very local ? |
| Stream |

J F M A M J J A S O N D

Overall length:	24–30mm
Hindwing:	14–18mm

♂ pronotum: arched, without distinct central lobe

COMPARISON OF WINGSPOTS

< 2× as long as wide

PERSIAN BLUETAIL

≥ 2× as long as wide

COMMON BLUETAIL

This Middle Eastern species was discovered in Cyprus in 2013, where it is now known to occur in five valleys in the west of the island. Although its European conservation status has yet to be evaluated, it is very likely to be Endangered (globally, it is currently known from just 12 localities and is categorized as Near-threatened). Very similar to Small Bluetail, but not known to occur in the same location as that species.

Identification: General appearance like Small Bluetail (*p. 102*): forewings of both sexes have **two-toned**, diamond-shaped wing-spots, **shorter** than those of Common Bluetail (*p. 106*), < 2× as long as wide, and noticeably larger than those on hindwings; rear edge of pronotum arched, lacking a distinct central lobe. ♂ Essentially black-and-blue; **mostly blue S8–9**, with variable black markings; wing-spots on forewing larger than those on hindwing. ♀ Olive-green side to thorax and very thin dark shoulder stripes; pronotum has two pale spots in centre; abdomen **tapering** on top of S8–10 (unlike Small Bluetail) and **lacking pale S8**; wing-spots browner and less clearly two-toned than male. VARIATION: Males of first generation have less blue on S8–9; immature females have orange on thorax and S1–2, like Small Bluetail.

Behaviour: Has at least two generations per year, but disappears from sites that dry out in summer.

Breeding habitat: Permanent, slow-flowing streams and channels, with localized patches of reeds or other marshy vegetation.

LOOK-ALIKES (in range)
Common Bluetail (*p. 106*)

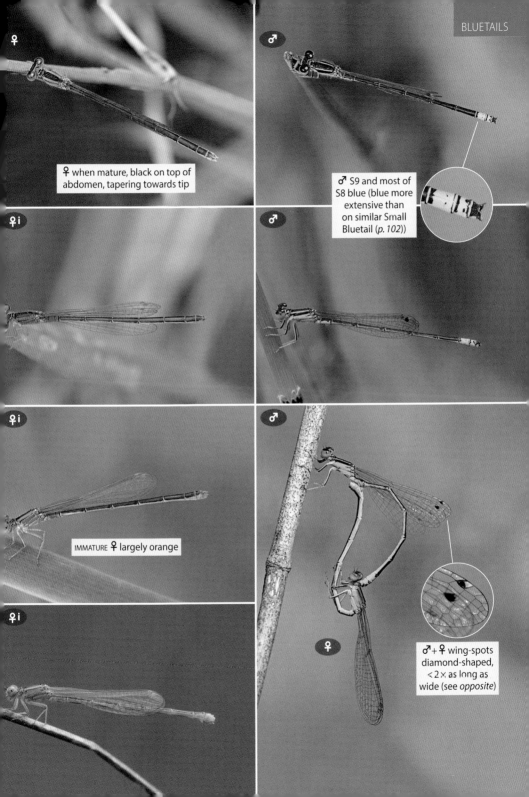

♀ when mature, black on top of abdomen, tapering towards tip

♂ S9 and most of S8 blue (blue more extensive than on similar Small Bluetail (*p. 102*))

IMMATURE ♀ largely orange

♂ + ♀ wing-spots diamond-shaped, < 2 × as long as wide (see *opposite*)

LC Common Bluetail

Ischnura elegans ×2

Blue-tailed Damselfly, Common Bluetip

Very common	=

Lake, pond, river, stream, canal, ditch

J F M A M J J A S O N D

Overall length:	30–34 mm
Hindwing:	14–20 mm

♂ pronotum: upright projection usually a distinct 'spike'

♂ + ♀ wing-spot: ≥ 2× as long as wide

One of the commonest dragonflies, easily found in a wide variety of habitats, except in N and SW Europe; it is replaced in the SW by other, almost identical, bluetails.

Identification: Like most other bluetails, both sexes have **two-toned**, diamond-shaped wing-spots on forewings, whitish towards tip and ≥ 2× as long as wide; abdomen mainly black on top, apart from blue **S8** (brown on some females); pronotum usually with distinct **upright projection** ('spike') in the middle of straight rear edge. ♂ Base colour blue, including narrow shoulder stripes and S8; viewed from behind, tips of inner branches of upper appendages parallel. ♀ Three different **colour forms** when mature: male-like; with olive-green thorax and dark brown S8; and pale brown with dull brown S8. Wing-spots are less clearly patterned than on male. **VARIATION:** Pronotum 'spike' absent on some females, and both sexes in SE Europe, where inner branches of male upper appendages may cross. Thorax greenish on immature males. Immature females have violet side to thorax (maturing to either male-like coloration or olive-green thorax and dark brown S8), or have orange-pink side to thorax and indistinct dark lines below broad, pale shoulder stripes (maturing to pale brown with dull brown S8).

Behaviour: Most adults remain close to water, often congregating in large numbers in marginal vegetation. May mate before attaining mature coloration. Mating prolonged (up to six hours) and typically peaks in afternoon. Eggs are laid into plants and debris at the surface, without males in attendance. Jerks rapidly up and down in flight when interacting with other damselflies.

Breeding habitat: Wide range of mainly lowland habitats, but uncommon in acidic and fast-flowing waters. Tolerates brackish and polluted water.

LOOK-ALIKES (in range)
Other bluetails (*pp. 102–119*)
Redeyes (*pp. 98–101*)

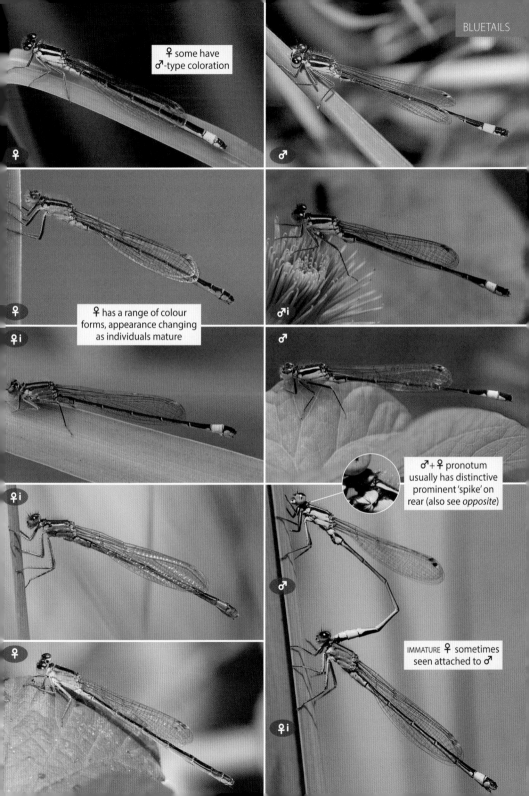

♀ some have ♂-type coloration

♀ has a range of colour forms, appearance changing as individuals mature

♂ + ♀ pronotum usually has distinctive prominent 'spike' on rear (also see *opposite*)

IMMATURE ♀ sometimes seen attached to ♂

♀

♀

♀ i

♀ i

♀

♂

♂ i

♂

♂

♀ i

LC Iberian Bluetail

Ischnura graellsii ×2

Common	=

Any standing or flowing water

J F M A M J J A S O N D

Overall length:	26–31 mm
Hindwing:	13–19 mm

♂ pronotum: arched, without distinct upright 'spike'

♂ + ♀ wing-spot: ≥ 2 × as long as wide

This Common Bluetail look-alike is endemic to the Iberian Peninsula and NW Africa and generally replaces it over much of Iberia, where it can be very common. It may hybridize with Common Bluetail (*p. 106*) where both species occur. Some females may not be identified safely, even in the hand.

Identification: Confirming identification requires in-hand examination of the male appendages. Typically slightly smaller than almost identical Common Bluetail, about the same size as Small Bluetail (*p. 102*); **lacks** prominent upright 'spike' on rear edge of pronotum usually shown by Common Bluetail; eye-spots often small; pale shoulder stripes vary in width, absent on some. ♂ Tips of inner branches of upper appendages **diverge** and lower appendages **curve inwards** slightly (parallel and curve outwards, respectively, on Common Bluetail). ♀ Typically has black markings on otherwise blue or brownish S8; pink/red forms of female rare. VARIATION: Immature females greenish-yellow, pale lilac or orange, last with S1–2 orange, like Small Bluetail.

Behaviour: Most adults remain close to water, often congregating in marginal vegetation. Has at least two generations per year.

Breeding habitat: As with Common Bluetail, a wide range of wetlands may be occupied, including brackish waters.

LOOK-ALIKES (in range)
Common Bluetail (*p. 106*)
Small Bluetail (*p. 102*)

♀i

♂

♀ often black marks on S8

♀

♂

♀i ♂+♀ eye-spots and shoulder stripes often restricted or even absent

♂ ♂+♀ pronotum lacks 'spike' on rear (usually shown by Common Bluetail)

♀ has a range of colour forms, appearance changing as individuals mature

♀

♀i

♀

♀

E LC Island Bluetail

Ischnura genei ×2

| Common | = |
| Standing and flowing waters | |

J F M A M J J A S O N D

Overall length:	29–32 mm
Hindwing:	12–18 mm

♂ pronotum: central lobe weakly divided

Endemic to the central Mediterranean, replacing Common Bluetail in the Tuscan archipelago and the Tyrrhenian and Maltese islands, except on Elba and Giglio, Tuscany, where both species occurred prior to 1990. Identification should ideally be confirmed in the hand, especially where Common Bluetail (*p. 106*) is known to occur.

Identification: Like Common Bluetail, both sexes have two-toned, diamond-shaped wing-spots on forewings, whitish towards tip and at least twice as long as wide; abdomen mainly black on top, apart from blue S8 (brown on some females). The only characters separating this species from Common Bluetail are the **short, weakly divided lobe** in the centre of the rear edge of the pronotum and **green** 'face', thorax and base of abdomen; wing-spots equal in size, unlike Small Bluetail (*p. 102*). ♂ As Common Bluetail, but 'face', thorax and base of abdomen **green**; inner branches of upper appendages **cross over** at the tips. ♀ Occurs in various colour forms, much like Common Bluetail but thorax never blue; pale on S8 always **constricted** by black at each end. VARIATION: Green on males and some females turns turquoise when fully mature.

Behaviour: Most adults remain in and around marginal vegetation at breeding sites.

Breeding habitat: Occupies well-vegetated waters, both standing and slow-flowing.

LOOK-ALIKES (in range)
Common Bluetail (*p. 106*)
Small Bluetail (*p. 102*)

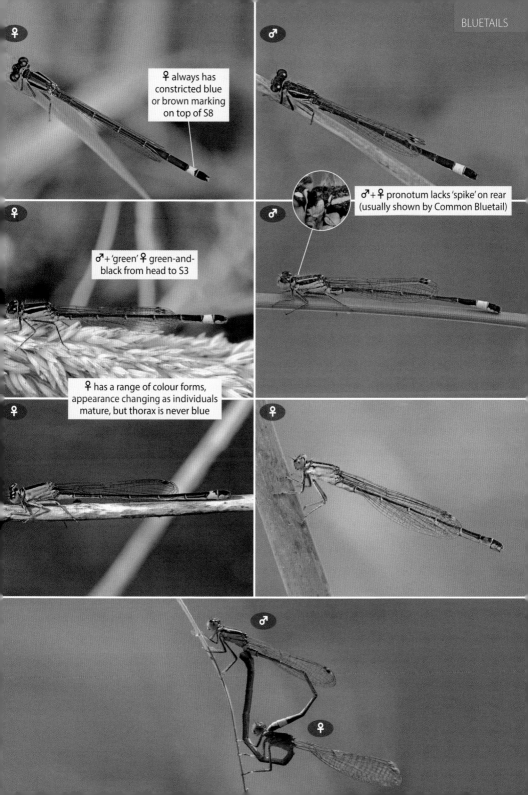

♀ always has constricted blue or brown marking on top of S8

♂ + ♀ pronotum lacks 'spike' on rear (usually shown by Common Bluetail)

♂ + 'green' ♀ green-and-black from head to S3

♀ has a range of colour forms, appearance changing as individuals mature, but thorax is never blue

LC Sahara Bluetail

Ischnura saharensis ×2

Common =

Any standing or flowing water

J F M A M J J A S O N D

Overall length:	26–31 mm
Hindwing:	12–17 mm

♂ pronotum: slightly arched, without distinct 'spike'

Widespread in the Canary Islands, occurring on all islands except Hierro. Very similar to Iberian Bluetail (*p. 108*), Island Bluetail (*p. 110*) and Marsh Bluetail (*p. 114*), but only the last of these occurs in the Canary Islands (on Tenerife and La Palma).

Identification: Small (same size as Small Bluetail (*p. 102*)) and distinctly orange on sides of S3–7, a feature shared with Marsh Bluetail; pronotum without upright 'spike'; wing-spots are all the same size. ♂ Sides of thorax and base of abdomen green or pale blue; **straight** black/green division on side of S2 (wavy bulge on Marsh Bluetail); projection on top of tip of S10 small but fairly obvious (tiny on Marsh Bluetail); inner branches of upper appendages **cross over** at the tips. ♀ Sides of female thorax and S1–2 various colours, some related to immaturity; may lack black below pale shoulder stripes; S8 blue or dull brownish (black on Marsh Bluetail). VARIATION: Pale shoulder stripes may be reduced or absent on males; sides of female thorax and S1–2 green (like male), blue, violet, orange-pink or dull green-brown (last two may lack black below pale shoulder stripes).

Behaviour: Has several generations throughout the year, although scarce in winter. Readily colonizes newly created wetland sites.

Breeding habitat: All types of standing and flowing waters, preferably with dense vegetation. Tolerant of brackish waters and can survive in ephemeral waters due to rapid larval development.

LOOK-ALIKES (in range)
Marsh Bluetail (*p. 114*)

♀

♂

♂ division between black and green on side of S2 side straight

♀ always has blue or dull brown S8 'tail light'

♀

♀

♂ + 'green' ♀ green-and-black from head to S3

♀

♀ i

♂ + ♀ side of abdomen distinctly orange

♀ has a range of colour forms, including orange, appearance changing as individuals mature

♀ i

♂

♀

NE Marsh Bluetail

Ischnura senegalensis ×2

Common Bluetail
Tropical Bluetail

Very local ?

Any standing or flowing water

J F M A M J J A S O N D

| Overall length: | 24–33 mm |
| Hindwing: | 14–16 mm |

♂ pronotum: without distinct upright 'spike'

Common and widespread in a wide range of habitats across much of Africa and Asia, Marsh Bluetail has only recently been found to occur on some of the Canary Islands (Tenerife and La Palma). Orange form females, presumed to have been this species, were found at a small reservoir in the extreme south of Tenerife in February 1993, but it was not until 2007 that DNA taken from a specimen on the island confirmed the species' presence.

Identification: Like most bluetails, both sexes have two-toned, diamond-shaped wing-spots on forewings, pale towards the tip and at least twice as long as wide; pronotum without upright 'spike'; wing-spots are all the same size. ♂ Abdomen mainly black on top with blue S8; **wavy edge** to black/green on side of S1–2; sides of S3–7 distinctly orange (like Sahara Bluetail (*p. 112*) and Oasis Bluetail (*p. 116*)); projection on top of tip of S10 insignificant. ♀ Dark shoulder stripes lacking; abdomen all-dark on top, **including S8**; wing-spots are less clearly patterned than on male. VARIATION: Shoulder stripes may be absent on males and sides of thorax and S1–2 blue; sides of female thorax and S1–2 may be green (like male), blue, **orange** (similar to immature Small Bluetail (*p. 102*)), cinnamon or olive.

Behaviour: As with other bluetails, often seen mating in waterside vegetation. Has a long flight period and may occur in all months.

Breeding habitat: All types of open standing and slow-flowing waters, including brackish and nutrient-enriched.

LOOK-ALIKES (in range)
Sahara Bluetail (*p. 112*)

♀

♂

♂ side of abdomen distinctly orange

♀

♂

♂ division between black and green on side of S2 side wavy

♀ i

♀ top of abdomen all-dark

Although very localized in Europe, occasionally this species has emerged from aquatic plants imported from Asia.

♂

♀ has a range of colour forms, including orange, appearance changing as individuals mature

♀ i

♀

ⓋⓊ Oasis Bluetail

Ischnura fountaineae ×2

| Rare and very local ❓ |
| Lake, pond, stream |

J F M A M J J A S O N D

| Overall length: | 27–34 mm |
| Hindwing: | 19–24 mm |

♂ pronotum: without distinct upright 'spike'

This desert species breeds on the Italian island of Pantelleria, between Sicily and Tunisia, in Specchio di Venere, a hot, sulphurous crater lake. There is also a single record from the nearby island of Linosa.

Identification: Relatively large, similar in size to Island Bluetail (*p. 110*) with small eye-spots; pale shoulder stripes narrow or absent; orange sides to S3–7 (like Marsh Bluetail (*p. 114*) and Sahara Bluetail (*p. 112*), which, in Europe, are found only on the Canary Islands); pronotum without upright 'spike'. ♂ Sides of thorax and S1–2 **blue**, the latter with **wavy edge** to black/blue division; appendages all the same length; tips of inner branches of upper appendages not crossed. ♀ Usually lacks dark shoulder stripe and pale S8 'tail light'; base colour olive to brown. VARIATION: Some females have blue sides to thorax and S1–2 and blue 'tail light' on S8; immature males whitish (green on other bluetails); immature females lack shoulder stripes and have orange sides of thorax and S1–2, extending up side of S8 on some.

Behaviour: Although it has two generations and flies all year in N Africa, there is probably only one generation on Pantelleria.

Breeding habitat: In addition to the hot spring-fed lake at Specchio di Venere, in N Africa it is also found in coastal wetlands and desert oases and wadis. The larvae can tolerate extreme conditions, notably water temperatures of around 50°C and salinity of up to 2·3%.

| LOOK-ALIKES (in range) |
| Other bluetails (*pp. 102–115*) [but none occurs on Pantelleria] |

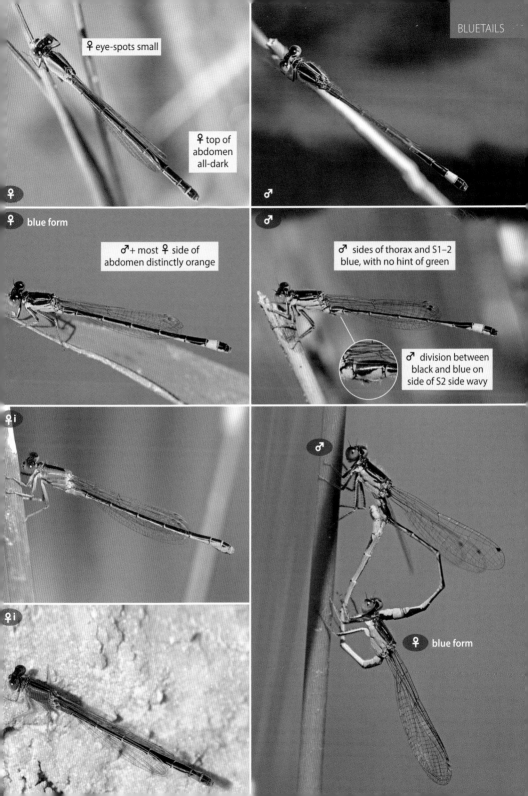

♀ eye-spots small

♀ top of abdomen all-dark

♀

♂

♀ blue form

♂

♂+ most ♀ side of abdomen distinctly orange

♂ sides of thorax and S1–2 blue, with no hint of green

♂ division between black and blue on side of S2 side wavy

♀ i

♀ i

♂

♀ blue form

VU Citrine Forktail

Ischnura hastata ×2

Locally common ▽

Pond, lake

J F M A M J J A S O N D

Overall length:	20–27 mm
Hindwing:	11–15 mm

This highly dispersive member of the bluetail genus from the Americas has a unique population in the Azores: it is the only parthenogenic dragonfly population known anywhere in the world, females breeding here in the absence of males.

Identification: Smallest European damselfly. ♂ (not recorded in Europe): thorax black and pale green; abdomen largely **yellow**, marked in black on S1–7/8, with conspicuous **spike** on S10; uniquely, **orange** wing-spot on forewing **does not reach** front edge of wing, where vein is **thickened and white**. ♀ Olive or grey with top of abdomen dark when mature, some **becoming pruinose** greyish when mature; wing-spots uniform grey-brown, all the same size. VARIATION: Immature females have orange body with black markings; abdomen largely unmarked on S1–4 (only S1–2 on orange form of Small Bluetail (*p. 102*)).

Behaviour: In N America, females mate only once and may then be dispersed by the wind; it is possible that such individuals might be carried on strong winds across the Atlantic and may be found away from the known population in the Azores. Its dispersive behaviour has also led to it being the only damselfly present in the Galápagos archipelago in the Pacific Ocean.

Breeding habitat: Well-vegetated ponds and lakes with pondweed (*Potamogeton* spp.) and spike-rush (*Eleocharis* spp.) from sea level to 850 m; also tolerates temporary and brackish water, at least in N America.

LOOK-ALIKES (in range)
Small Bluetail (*p. 102*)

♀

♂

♂ unmistakable, although not known in the
only European population (in the Azores)

♀

♂

♀i

♀

♀ mainly orange at first, dull when
mature; may become pruinose

A mating pair photographed in the USA;
mating has not been recorded in Europe.

♂

♀

Male 'blue-tailed' damselflies compared *p. 95*

NT Sedgling

Nehalennia speciosa ×2

Pygmy Damselfly

| Scarce and local ▽ |
| Bog, lake |

J F M A M J J A S O N D

Overall length:	24–26 mm
Hindwing:	11–16 mm

This tiny, unobtrusive, green and blue species is a highly localized denizen of bogs and lakesides, especially in the Bavarian Prealps, northern Poland and the Baltic States. Its range has contracted markedly since the 1960s as a result of environmental factors. One of only six species in its genus, the other five inhabiting the Americas, where they are known as sprites.

Identification: Smallest damselfly on European mainland; both sexes lack eye-spots and have **metallic** green upperparts (bronzy with age); wings very short, only two-thirds length of abdomen; wing-spots small and pale. ♂ **Pale blue** eyes, arc across back of head, tapering 'wedge'-shape on S8–9 and **S10**; tip of S10 has pair of tiny spines on top; upper appendages relatively large and blunt. ♀ Coloured as male, but blue areas turn brown with age.

Behaviour: As befits a tiny species, spends much of its time perching or flying weakly within sedge beds or similar cover.

Breeding habitat: Very specific habitat requirements, breeding in shallow acidic waters of peat bogs or lake fringes. These have good cover provided by beds of sedges (*Carex* spp.), or exceptionally other species such as Purple Moor-grass (*Molinia caerulea*) or Water Horsetail (*Equisetum fluviatile*). Has poor dispersal ability, breeding in small, well-scattered colonies. Declining and now extinct in many W parts of its range. Breeding sites lost to drainage and climate change or threatened by eutrophication and wetland succession.

LOOK-ALIKES (in range)
Spreadwings (*pp. 20–33*)
Small Bluetail (*p. 102*)

120

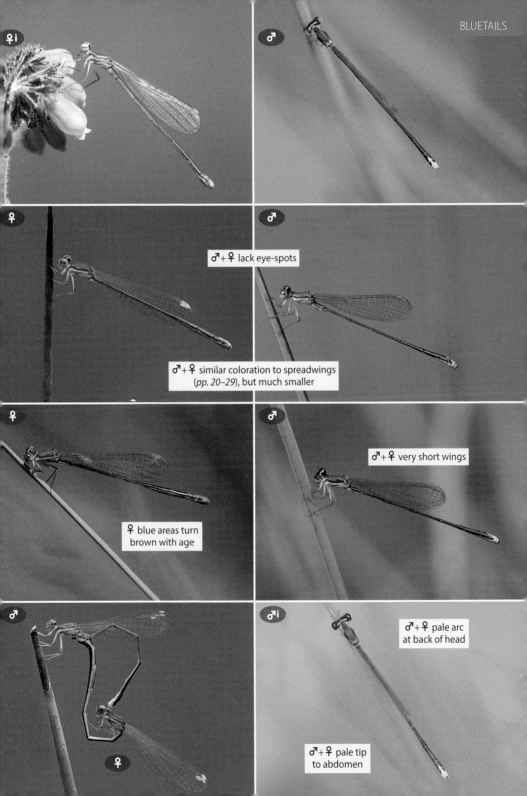

♀i

♂

♀

♂

♂ + ♀ lack eye-spots

♂ + ♀ similar coloration to spreadwings
(pp. 20–29), but much smaller

♀

♂

♂ + ♀ very short wings

♀ blue areas turn
brown with age

♂

♀

♂i

♂ + ♀ pale arc
at back of head

♂ + ♀ pale tip
to abdomen

Introduction to dragonflies (Anisoptera)

Dragonflies are typically larger than damselflies, with a robust structure and more powerful flight. The wings are held out perpendicular to the body when at rest and the hindwings are broader at the base than the forewings. The eyes touch either broadly or narrowly, or are widely separated. In some groups, individuals sit more-or-less horizontally on territorial perches ('perchers'), whereas the others fly for long periods over their wetland territories ('fliers') and are only occasionally seen at rest, hanging vertically. These characteristics are similar for females and for males when away from water.

The 93 dragonflies can be grouped into nine basic 'types', based on their general appearance. The broad characteristics of each are summarized here.

HAWKERS, SPECTRES & EMPERORS (p. 126)
GENERA: *Brachytron, Aeshna, Caliaeschna, Boyeria* & *Anax*   20 species

Hawkers (p. 127)
GENERA: *Brachytron* & *Aeshna*   12 species

Large to very large 'fliers'; eyes meet broadly; distinctive and usually prominent pale markings on side of thorax; long, thin abdomen either has pairs of blue, green or yellow dots along length ('mosaic' hawkers) or (two species) is largely brown; powerful flight; territorial males often patrol over water, generally below waist height; typically feed away from water; hang more-or-less vertically when perched; usually standing waters; eggs laid at rest alone into plant material or debris.

Hawker

Spectres (or dusk-hawkers) (p. 127)
GENERA: *Caliaeschna* & *Boyeria*   3 species

Large hawker-like 'fliers'; eyes meet broadly; thorax and long, thin abdomen well patterned, latter with pale tip ('tail light'); territorial males patrol low along shady streams and rivers in S Europe; hang more-or-less vertically when perched; eggs laid at rest alone into marginal debris and mosses.

Spectre

Emperors (p. 127)
GENUS: *Anax*   5 species

Large to very large 'fliers'; eyes meet broadly; 4 widespread species have virtually unmarked thorax and blue base to long, thin abdomen; hang more-or-less vertically when perched; territorial males patrol over water for long periods, often above waist height; may undertake long migrations; mainly standing waters; eggs laid at rest into vegetation, alone or in tandem.

Emperor

CLUBTAILS, SNAKETAIL, PINCERTAILS, HOOKTAIL & BLADETAIL (p. 176)
GENERA: *Gomphus, Stylurus, Ophiogomphus, Onychogomphus, Paragomphus & Lindenia*

15 species

Medium-sized to large 'perchers'; eyes do not touch; complex, mostly black and yellow, body markings; male abdomen may have swollen tip and prominent curved appendages; territorial males perch horizontally on trees or ground; most breed in running waters, but often found well away from water; eggs laid alone in flight into water.

Clubtail

Pincertail

GOLDENRINGS, CRUISER & CASCADER
GENERA: *Cordulegaster, Macromia & Zygonyx*

9 species

Goldenrings (or golden-ringed dragonflies) (p. 209)
GENUS: *Cordulegaster* 7 species

Very large 'fliers' (longest European species); green or blue-green eyes just touch; body black with bold yellow stripes and bands, including bands across the long abdomen; ovipositor sticks out beyond tip; males patrol low along flowing waters, often landing in bankside vegetation; perch more-or-less vertically, often to eat; eggs laid alone in flight with vertical jabs into substrate.

Goldenring

Cruiser (p. 209)
GENUS: *Macromia* 1 species

Large 'flier' with green eyes that meet broadly; black body with yellow stripes and bands; males fly persistently over slow-flowing waters or lakes in SW Europe; perches more-or-less vertically; eggs laid alone in flight.

Cruiser

Cascader (p. 209)
GENUS: *Zygonyx* 1 species

Medium-sized 'flier' with brown eyes that meet broadly; body black with yellow markings on side; males fly persistently over turbulent running waters in extreme SW Europe; eggs laid at rest alone, or in flight alone or in tandem.

Cascader

Introduction to dragonflies (Anisoptera) (continued)

EMERALDS (p. 230)

GENERA: *Cordulia, Corduliochlora, Somatochlora & Oxygastra*

9 species

Medium-sized 'fliers'; green eyes that meet broadly; body metallic bronzy/green with inconspicuous yellow markings; usually seen in flight patrolling regular beats along margins of standing or flowing waters or over boggy areas; most species feed around canopy of trees, where they land to feed; eggs laid in flight alone into water or marginal debris.

Emerald

BASKETTAIL & CHASERS (p. 250)

GENERA: *Epitheca & Libellula*

4 species

Baskettail (p. 250)

GENUS: *Epitheca*

1 species

Large 'flier'; eyes dull blue-green, meeting broadly; dark body; dark markings at hindwing bases; strong fliers, cruising high over standing waters and feeding areas; rarely perch; eggs laid in flight into water, alone, from 'basket' under tip of abdomen.

Baskettail

Chasers (p. 252)

GENUS: *Libellula*

3 species

Medium-sized 'perchers'; eyes meet broadly; abdomen pruinose blue or brownish; hindwing bases dark (not always easy to see); perch horizontally, flying up to catch food, chase off intruders or pursue females, often returning to the same perch; mates quickly; eggs laid in flight into water, alone but usually with male in close attendance.

Chaser

SKIMMERS (p. 258)

GENUS: *Orthetrum*

9 species

Small to medium 'perchers'; abdomen pruinose blue (males) or yellowish-brown (females and immatures); hindwing bases clear; antenodal cross-veins yellow; often skim low over water; perch on ground or among low vegetation; eggs laid in flight into water, usually alone with male in close attendance.

Skimmer

WHITEFACES (p. 282)

GENUS: *Leucorrhinia*

5 species

Small 'perchers'; 'face' white; body red- or pruinose blue-and-black (mature males) or yellow-and-black (females and immatures); hindwing bases dark; perch low down and behave much like darters; mostly N Europe; eggs laid in flight alone, but with males in attendance.

Whiteface

DARTERS, SCARLET, DROPWINGS, PENNANT, PERCHER & GROUNDLING (p. 294)

GENERA: *Sympetrum, Crocothemis, Trithemis, Selysiothemis, Diplacodes & Brachythemis*   19 species

Darter

Darters & Scarlet (p. 294)

GENERA: *Sympetrum & Crocothemis*   12 species

Small 'perchers'; body red or black (mature males) or yellowish-brown (females and immatures); hindwing bases clear; sit typically on bare surfaces or prominent perches, flying up to catch food, chase off intruders or pursue females, often returning to same perch; eggs laid in flight into standing waters, often in tandem.

Dropwings (p. 295)

GENUS: *Trithemis*   4 species

Small 'perchers'; mature males violet, red or dark; females and immatures yellowish, with dark, diagonal markings across side of thorax; perch on or near the ground; standing or flowing waters; eggs laid in flight into water, by female alone. S Europe.

Dropwing

Pennant & Percher (p. 295)

GENERA: *Selysiothemis & Diplacodes*   2 species

Very small 'perchers'; black (mature males) or beige (females and immatures); perch near ground at standing waters in southern Europe; eggs laid in flight into water, typically in tandem.

Percher

Groundling (p. 295)

GENUS: *Brachythemis*   1 species

Small 'percher'; body black (male) or beige (female and immatures); wing-spots two-toned; mature males and some females have conspicuous black wing bands; sit on bare ground, sometimes in large numbers, in SW Europe; lays eggs in flight into water.

Groundling

GLIDERS (p. 338)

GENERA: *Tramea & Pantala*   2 species

Medium-sized 'fliers'; abdomen yellow or red, often angled down; long, broad-based wings; wing spots shorter in hindwing; flies persistently, often above head height; perches vertically; tandem pairs lay eggs into water; regularly make long migrations; rare in Europe, most regular in Cyprus.

Glider

HAWKERS, SPECTRES & EMPERORS

FAMILY | **Aeshnidae**

GENERA | *Brachytron, Aeshna, Caliaeschna, Boyeria & Anax*

20 SPECIES

Large to very large (mean lengths 59–83 mm) persistent 'fliers' with eyes that meet broadly. Robust, with a long, more-or-less parallel-sided abdomen. Bodies range from very colourful, through intricate patterning to cryptic brown. Males of most species are aggressively territorial, patrolling breeding waters for long periods (more often found at water than females), sometimes in aggregations, over open areas or along woodland edges. Hang more-or-less vertically when perched, either in waterside vegetation or on tall grasses, shrubs or trees some distance from water. Eggs laid, using prominent ovipositor, into organic debris or plant tissues, in most species by unaccompanied females. Breeding habitats varied; some habitat generalists are prone to moving long distances, facilitating recent range spread.

Key features of hawkers, spectres & emperors

◆ large–very large 'fliers'
◆ eyes meet broadly
◆ long, thin abdomen
◆ hang more-or-less vertically
◆ prominent ovipositor
◆ most are aggressively territorial
◆ may range far from water

Identify 'sub-types' by colour/patterning of thorax and abdomen | colour of eyes and costa | suffusion in wings | colour, size and shape of wing-spots | range

Key identification features

BLUE HAWKER ♂
Aeshna cyanea

EYE: colour

COSTA: colour

THORAX: colour and patterning on side

THORAX: presence and width of pale shoulder stripe or overall colour

S2: markings

WING-SPOT: colour and shape

WING: colour

ABDOMEN: extent and patterning of coloured markings

S9–10: markings

Pointers on the images shown on the comparison plates (*pp. 128–131*) indicate the key identification features to focus on when viewing each species from above. Full details are given in the relevant species account.

Hawkers   12 species
GENERA: *Brachytron (p. 136)* & *Aesha (pp. 138–159)*

KEY FEATURES:
- thorax side has yellow or green stripes
- thorax top often has yellow stripes
- abdomen either black with coloured dots or mostly brown
- breed mainly in standing waters
- territorial males fly mostly below 1 m
- feed high late in day away from water
- eggs laid at rest into marginal debris

Identify species by pattern of colour on top and side of thorax | extent and colour of markings on abdomen | colour of eyes and costa | suffusion in wings

Hawker

Spectres   3 species
GENERA: *Caliaeschna (p. 160)* & *Boyeria (pp. 162–165)*

KEY FEATURES:
- size of small hawkers
- shoulder stripes hooked at rear
- 'tail light' across S9 and/or S10
- breed in flowing waters
- territorial males fly low in shade
- eggs laid at rest into mosses and roots on banks
- active late in day, attracted to lights

Identify species by
pattern on thorax and abdomen | range

Spectre

Emperors   5 species
GENUS: *Anax (pp. 166–175)*

KEY FEATURES:
- thorax plain green or brown, or with bold pale stripes on side
- male abdomen with blue base at least
- breed mainly in standing waters
- males patrol for long periods, often above 1 m
- eggs laid at rest into plants, in tandem in two species
- fly strongly, land infrequently, feed in flight
- some undertake long migrations

Identify species by colour of eyes, thorax and abdomen

Emperor

Male hawkers compared

● = widespread; **N** = northern and/or montane; **C** = central;
S = southern; **SE** = south-eastern; **SW** = south-western; **v** = vagrant.

N/C	●	C/S	●	N	N
Hairy Hawker *Brachytron pratense* (p. 136)	**Migrant Hawker** *Aeshna mixta* (p. 138)	**Blue-eyed Hawker** *Aeshna affinis* (p. 140)	**Blue Hawker** *Aeshna cyanea* (p. 142)	**Green Hawker** *Aeshna viridis* (p. 144)	**Azure Hawker** *Aeshna caerulea* (p. 146)
N/C	N	N	N	N/C	C/S
Moorland Hawker *Aeshna juncea* (p. 148)	**Bog Hawker** *Aeshna subarctica* (p. 150)	**Siberian Hawker** *Aeshna crenata* (p. 152)	**Baltic Hawker** *Aeshna serrata* (p. 154)	**Brown Hawker** *Aeshna grandis* (p. 156)	**Green-eyed Hawker** *Aeshna isoceles* (p. 158)

Male spectres and emperors compared

Pointers indicate key identification features.

SW	**SE**	**SE**
Western Spectre *Boyeria irene* (p. 162)	**Cretan Spectre** *Boyeria cretensis* (p. 164)	**Eastern Spectre** *Caliaeschna microstigma* (p. 160)

●	**v**	●	**S**	**SE**
Blue Emperor *Anax imperator* p. 166	**Common Green Darner** *Anax junius* p. 168	**Lesser Emperor** *Anax parthenope* p. 170	**Vagrant Emperor** *Anax ephippiger* p. 172	**Magnificent Emperor** *Anax immaculifrons* p. 174

Female hawkers compared

● = widespread; **N** = northern and/or montane; **C** = central;
S = southern; **SE** = south-eastern; **SW** = south-western; **v** = vagrant.

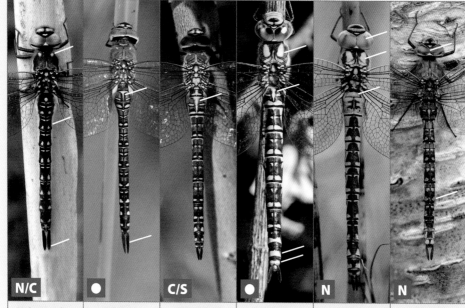

N/C

Hairy Hawker
Brachytron pratense
(p. 136)

●

Migrant Hawker
Aeshna mixta
(p. 138)

C/S

Blue-eyed Hawker
Aeshna affinis
(p. 140)

●

Blue Hawker
Aeshna cyanea
(p. 142)

N

Green Hawker
Aeshna viridis
(p. 144)

N

Azure Hawker
Aeshna caerulea
(p. 146)

N/C

Moorland Hawker
Aeshna juncea
(p. 148)

N

Bog Hawker
Aeshna subarctica
(p. 150)

N

Siberian Hawker
Aeshna crenata
(p. 152)

N

Baltic Hawker
Aeshna serrata
(p. 154)

N/C

Brown Hawker
Aeshna grandis
(p. 156)

C/S

Green-eyed Hawker
Aeshna isoceles
(p. 158)

Female spectres and emperors compared

Pointers indicate key identification features.

Western Spectre	**Cretan Spectre**	**Eastern Spectre**
Boyeria irene	*Boyeria cretensis*	*Caliaeschna microstigma*
(p. 162)	(p. 164)	(p. 160)

Blue Emperor	**Common Green Darner**	**Lesser Emperor**	**Vagrant Emperor**	**Magnificent Emperor**
Anax imperator	*Anax junius*	*Anax parthenope*	*Anax ephippiger*	*Anax immaculifrons*
p. 166	p. 168	p. 170	p. 172	p. 174

Male hawkers in flight

● = widespread; **N** = northern and/or montane;
C = central; **S** = southern.

Hairy Hawker
Brachytron pratense
(p. 136)

N/C

Small, usually low-flying; side of thorax pale; abdomen held more or less horizontal. Earliest emerging hawker.

Migrant Hawker
Aeshna mixta
(p. 138)

●

Broad pale stripes on side of thorax and faint markings on top; abdomen uptilted and slightly droop-tipped. Sometimes in aggregations.

Blue-eyed Hawker
Aeshna affinis
(p. 140)

C/S

Eyes all-blue; side of thorax mainly bluish; abdomen held rather straight.

Blue Hawker
Aeshna cyanea
(p. 142)

●

Thorax with very broad pale 'head-lights' on top and very pale side; abdomen held more or less horizontal and slightly downcurved, with 'tail-lights' (pale bands across S9 & 10). Often inquisitive.

Green Hawker
Aeshna viridis
(p. 144)

N

Eyes blue; thorax with broad green 'head-lights' on top and green side; abdomen with bold blue markings.

Azure Hawker
Aeshna caerulea
(p. 146)

N

Very narrow, wavy pale lines on side of thorax; abdomen shows more blue than black from side.

Hawkers are most often seen in flight; males of the 12 species are shown here at life-size; the key features to look for are described.

Moorland Hawker
Aeshna juncea
(p. 148)

N/C

Narrow pale stripes on top and side of thorax; abdomen uptilted and held straight; costa yellow.

Bog Hawker
Aeshna subarctica
(p. 150)

N

Narrow pale stripes on top and side of thorax; costa brown; black line across 'face' thick where meets eyes.

Siberian Hawker
Aeshna crenata
(p. 152)

N

Narrow pale stripes on top and bold stripes on side of thorax; markings on abdomen all-blue; costa brown.

Baltic Hawker
Aeshna serrata
(p. 154)

N

Bold pale stripes on top and sides of thorax; large spots on abdomen; costa yellow.

Brown Hawker
Aeshna grandis
(p. 156)

N/C

Amber-tinted wings; thorax brown with yellow stripes on side; abdomen brown with blue markings along side.

Green-eyed Hawker
Aeshna isoceles
(p. 158)

C/S

Eyes green; wings clear; thorax brown with yellow stripes on side; abdomen brown, unmarked on side.

Male spectres in flight

Spectres and emperors are most often seen in flight; males of the eight species are shown here at life-size; the key features to look for are described.

Western Spectre
Boyeria irene
(p. 162)

SW

Rather dull with obscure greenish markings; pale S9–10 forming indistinct 'tail-light'.

Cretan Spectre
Boyeria cretensis
(p. 164)

SE

Rather dull with obscure greenish markings; pale S9–10 forming indistinct 'tail-light'.

Eastern Spectre
Caliaeschna microstigma
(p. 160)

SE

Hooked rear end to stripes on top of thorax; broad pale stripes on side of thorax; abdomen rather dark; pale bands across S9 & 10 create 'tail-light'.

Male emporors in flight

● = widespread; **S** = southern; **SE** = south-eastern; **SW** = south-western;
v = vagrant.

Blue Emperor
Anax imperator
(p. 166)

Plain green thorax; abdomen often drooping, mainly blue side; yellow costa may be visible.

Common Green Darner
Anax junius
(p. 168)

v

As Blue Emperor, but blue on side of abdomen appears to taper slightly from base to tip (more parallel in Blue Emperor).

Lesser Emperor
Anax parthenope
(p. 170)

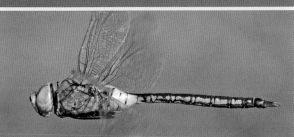

Plain brownish thorax; abdomen straight, with distinctive blue 'saddle' at the base; yellow costa and suffusion to outer half of wings may be visible.

Vagrant Emperor
Anax ephippiger
(p. 172)

S

Yellowish-brown with brown eyes; blue 'saddle' restricted to top of abdomen.

Magnificent Emperor
Anax immaculifrons
(p. 174)

SE

Blue eyes, mostly blue-green thorax/ S1 with black stripes on side; broad pale orange bands around abdomen.

LC **Hairy Hawker**

Brachytron pratense ×1

Hairy Dragonfly, Spring Hawker

| Locally common | = |

| Lake, pond, canal, ditch |

J F M A M J J A S O N D

| **Overall length:** | 54–63 mm |
| **Hindwing:** | 34–37 mm |

The smallest 'mosaic' hawker, flying in spring before the peak emergence of other hawkers. It is found rather sparingly at unpolluted waters with luxuriant marginal vegetation.

Identification: Small, rather dark 'mosaic' hawker with distinctive **hairy** thorax; best identified by combination of pair of relatively **small, oval-shaped** dots on the top of each abdominal segment, costa and **antenodal cross-veins yellow**, **long, thin, brown** wing-spots, and **long** appendages; side of thorax **extensively yellow-green**, lacking pattern of paired pale stripes characteristic of most other hawkers. ♂ Abdomen dark with small blue abdominal markings; strong yellow/green shoulder stripes; eyes blue; lacks distinctive acutely angled rear margin of hindwing typical of other male hawkers and spectres. ♀ As male, but shoulder stripes very restricted and abdomen with yellow markings and especially hairy. VARIATION: Immature coloration as female; rarely, female has yellow replaced with blue.

Behaviour: Males patrol territories at low level through and alongside emergent vegetation, seeking females and seeing off intruders. Mating generally takes place in vegetation close to water. Female rather secretive, usually only visiting water to find a mate or lay eggs. Eggs laid into floating decomposing vegetation, sometimes attended by male.

Breeding habitat: Ditches, canals, ponds and lakes with clean but standing or very slow-flowing water and abundant fringing tall emergent vegetation.

LOOK-ALIKES (in range)
Migrant Hawker (*p. 138*)
Blue-eyed Hawker (*p. 140*)
Azure Hawker (*p. 146*)
Eastern Spectre (*p. 160*)
Western Spectre (*p. 162*)

♂ ♂ + ♀ thorax extensively yellow-green

♂ earlier flight period than similar Migrant Hawker

♂ ♂ + ♀ thorax distinctly hairy

♂ ♂ + ♀ antenodal cross-veins yellow

♂ + ♀ complex 'mosaic' patterning

♂ + ♀ wing-spots long, thin, brown

♀

♀ ♂ + ♀ long appendages

LC Migrant Hawker

Aeshna mixta ×1

Hawkers compared *pp. 128–133*

Autumn Hawker

Common △

Lake, pond, river, canal, ditch

J F M A M J J A S O N D

Overall length:	56–64 mm
Hindwing:	37–41 mm

A relatively small, late-emerging 'mosaic' hawker flying well into autumn and often seen in large numbers. It is typically the commonest hawker and breeds widely across Europe, except in the north. Large migrations can occur, perhaps helping northward spread in recent years.

Identification: Thorax has yellow stripes on side; shoulder stripes short and weak or absent, abdomen has paired dots on each segment and **narrow yellow triangle** on S2; costa **brown**. In flight, the abdomen is held just above horizontal and appears slightly droop-tipped (see *p. 132*). The appendages are **very long**, at least as long as S9 & S10 combined. ♂ Appears quite dark, with pairs of blue dots and yellow flecks along abdomen; shoulder stripes **faint or absent**; eyes bluish. ♀ Brown, with similar abdominal pattern to male, but spots smaller and yellow; shoulder stripes very restricted or absent; wings very slightly tinted; eyes green-brown. VARIATION: Blue on male abdomen fades to violet in cool conditions; some females have blue spots on abdomen; immatures have blue areas replaced by yellow.

Behaviour: Males non-territorial, often flying low over water in some numbers, hovering frequently and investigating marginal vegetation in search of females. Mating prolonged, although female subsequently lays eggs alone into aquatic plants, often well above water. Both sexes frequently perch or hunt high up along hedgerows or woodland rides well away from water and fly until dusk, sometimes in 'swarms'. Often basks low down on vegetation.

Breeding habitat: Breeds in a variety of standing and slow-flowing waters edged with tall emergents; avoids acidic waters but tolerates brackish conditions.

LOOK-ALIKES (in range)

Hairy Hawker (*p. 136*)
Blue-eyed Hawker (*p. 140*)
Azure Hawker (*p. 146*)
Moorland Hawker (*p. 148*)
Bog Hawker (*p. 150*)
Spectres (*pp. 160–165*)

♂

Small, widespread hawker emerging late in the season; non-territorial and may form feeding or migratory swarms.

♂

♂+♀ narrow yellow triangle on S2

♂+♀ thorax with bold pale stripes on side

♀

♂

♂+♀ shoulder stripes indistinct or absent

♀

♂+♀ costa brown

♂+♀ very long appendages

LC Blue-eyed Hawker

Aeshna affinis ×1

Southern Migrant Hawker

Locally fairly common △

Pond, ditch

NORTH
SOUTH
J F M A M J J A S O N D

| **Overall length:** | 57–66 mm |
| **Hindwing:** | 37–42 mm |

An especially colourful small hawker with stunning blue eyes (male only) that flies somewhat earlier in the year than the similar Migrant Hawker (*p. 138*). This species has expanded its range northwards across continental Europe and now occurs several hundred kilometres farther N than in the 1990s.

Identification: Rather small 'mosaic' hawker similar to Migrant Hawker, but both sexes have side of thorax essentially pale with only **fine** black lines and lack dark central panel typical of most other 'mosaic' hawkers; top of thorax brown with short shoulder stripes; wings clear; costa pale; wing-spots ochre-coloured and slightly longer than those of Migrant Hawker. ♂ Side of thorax **green, becoming blue** with age; abdomen dark with pairs of **large** bright blue dots along length and **lacking** green or yellow markings when mature; **blue triangle ('golf tee')** shape on S2; eyes **bright blue**. ♀ Similar to male but with yellowish side to thorax and smaller paired yellow dots on abdomen; prominent **yellow triangle ('golf tee')** shape on S2; eyes olive/brown. VARIATION: As with other 'mosaic' hawkers, a rare blue form occurs in female.

Behaviour: Males patrol over drier areas away from open water at about 1·5 m – much higher than Hairy Hawker (*p. 136*). Flight weaker than that of Migrant Hawker, with more frequent hovering. Somewhat dispersive, at least in some years. The only 'mosaic' hawker that lays eggs in tandem.

Breeding habitat: Breeds in standing, sometimes slow-flowing or slightly brackish, waters with luxuriant emergent vegetation that often dry up during the summer.

LOOK-ALIKES (in range)
Hairy Hawker (*p. 136*)
Migrant Hawker (*p. 138*)
Green Hawker (*p. 144*)
Azure Hawker (*p. 146*)
Spectres (*pp. 160–165*)

♀

♀ blue form

♂+♀ pale triangle
('golf tee' shape) on S2

♂

♂ eyes bright blue

♀

blue form ♀

♂

♂

♂+♀ thorax
side pale with
fine black lines;
no dark panel
(unlike most other
'mosaic' hawkers)

♂ when mature,
conspicuously
blue from side,
including side
of thorax (see
Migrant Hawker
(p. 138))

LC Blue Hawker

Aeshna cyanea ×1

Southern Hawker

Common =

Lake, pond, canal, ditch

J F M A M J J A S O N D

Overall length:	67–76 mm
Hindwing:	43–51 mm

A large, solitary, inquisitive and colourful 'mosaic hawker'. It has spread N in recent years, but may be affected by climate change in the S of its range, where it lives mostly in mountainous areas.

Identification: Both sexes have **very broad**, coloured stripes on top and side of thorax and coloured **bands** across S9–10 (paired spots in most other 'mosaic' hawkers), both features often visible in flight (see *below*); **narrow yellow triangle** (shape of 'golf tee') on S2; wing-spots **short** (3 mm). ♂ Blackish with **apple-green** patterning, except for **sky blue** markings on S8–10 and side of abdomen; eyes bluish. ♀ Chocolate-brown with green markings. VARIATION: Female rarely has blue markings on all abdominal segments; immatures chocolate brown with pale yellow markings and brown eyes.

Behaviour: Adults mature away from water, often in woodland clearings, where they hawk insects up to size of butterflies, sometimes with other hawkers; feeding may continue until dusk on warm evenings. Mature males defend territories aggressively, flying above the water at a height of about 1 m. Females lay eggs into rotting vegetation, including timber, typically above the waterline; like other hawkers, the rustle of wings often indicates their presence.

Breeding habitat: Usually found in neutral or alkaline standing waters, often partly shaded by trees, and sometimes in slow-flowing waters. Found in garden ponds (where egg-laying females often predated by cats) and may be the only species at shady woodland ponds with many dead leaves.

♂ + ♀ show 'headlights' and 'tail lights' in flight

LOOK-ALIKES (in range)
Other 'mosaic' hawkers
(*pp. 136–155*), especially
Green Hawker (*p. 144*)
Western Spectre (*p. 162*)

♂

♂

♂ one of the most colourful of the hawkers, with extensive blue and apple-green markings

♂+♀ complete pale bands across S9–10

♂+♀ costa dark

♀ blue form

♂+♀ wing-spots short

♂+♀ thorax with very broad coloured bands on top and side

♀

♂+♀ narrow yellow triangle ('golf tee' shape) on S2

♀

NT Green Hawker

Aeshna viridis ×1

PROTECTED

Scarce and local ▽

Lake, canal, ditch

J F M A M J J A S O N D

| Overall length: | 65–75 mm |
| Hindwing: | 38–45 mm |

A large, colourful 'mosaic' hawker with striking similarities to Blue Hawker (*p. 142*), Blue-eyed Hawker (*p. 140*) and Blue Emperor (*p. 166*). The breeding habitat in its E European range comprises beds of Water-soldier *Stratiotes aloides*.

Identification: Slightly smaller than Blue Hawker; very **broad green** stripes on top of thorax; side of thorax **extensively green** with very limited black markings; costa **yellow**; **very thin** stem to black 'T'-mark on top of frons ('face'); outer part of wings **tinged yellowish**. ♂ Thorax dark brown with prominent apple-green shoulder stripes and side; abdomen also green at base, otherwise black with pairs of large blue spots and smaller yellow markings on top, side with much blue; eyes pale blue. ♀ Brown, with patterning similar to male but **all markings green** and with smaller pale markings on top of abdomen; eyes greenish. VARIATION: Immatures look like females.

Behaviour: Males patrol territories over areas dominated by Water-soldier, into which females lay eggs; spines on the leaves may damage wings. Both sexes may feed in groups away from water and rest low down in rough vegetation, where they may be hard to see.

Breeding habitat: Wetlands with extensive beds of Water-soldier, including oxbow lakes in floodplains, but mainly in artificial peaty ditches, canals and ponds, where rotational clearance is required to maintain suitable breeding conditions. Found locally in NE Europe and more rarely farther S in CE Europe.

LOOK-ALIKES (in range)

Blue-eyed Hawker (*p. 140*)
Blue Hawker (*p. 142*)
Siberian Hawker (*p. 152*)
Baltic Hawker (*p. 154*)
Blue Emperor (*p. 166*)

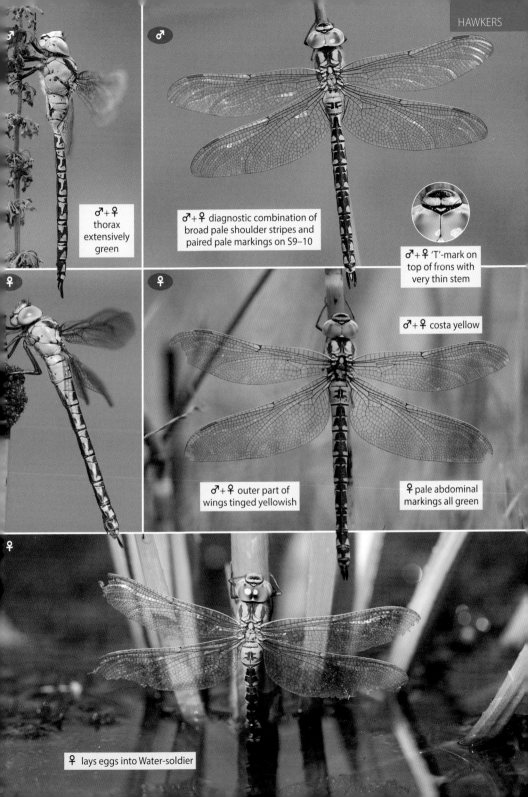

♂+♀ thorax extensively green

♂ ♂+♀ diagnostic combination of broad pale shoulder stripes and paired pale markings on S9–10

♂+♀ 'T'-mark on top of frons with very thin stem

♂+♀ costa yellow

♀ pale abdominal markings all green

♀ ♂+♀ outer part of wings tinged yellowish

♀ lays eggs into Water-soldier

LC Azure Hawker

Aeshna caerulea × 1

Common (N),
very local (C) ▽

Bog, pond

SCOTLAND
REST OF EUROPE
J F M A M J J A S O N D

| Overall length: | 54–64 mm |
| Hindwing: | 38–41 mm |

This is the smallest of a suite of mainly boreo-alpine 'mosaic' hawkers. Breeding at bog pools in N Europe and, farther S, mainly in the Alps, there is no overlap in breeding habitat with Blue-eyed Hawker (*p. 149*), which also appears very blue. An isolated population in Scotland flies a month earlier in the season than elsewhere.

Identification: Smaller than other hawkers with similar range and habitat; head relatively small; eyes meet for only **short** distance; pale stripe on side of thorax **wavy and very narrow**; costa **brownish**. ♂ Abdomen mostly blue, with large, paired markings and no yellow; pale shoulder stripes very short; eyes blue above, brown below. ♀ Brown, with similar abdominal pattern to male but all markings yellow; pale shoulder stripes very short or absent; eyes brownish. **VARIATION:** Blue and brown forms of female equally frequent, the latter and immatures having yellowish markings; blue fades to grey in cool conditions.

Behaviour: Active in sunny weather, males flying low over bog pools and surrounding moorland in search of females. Characteristically, in cool conditions both sexes bask on pale surfaces in sheltered locations. Mating occurs away from water; eggs are laid alone into soft substrates or bog-mosses (*Sphagnum* spp.).

Breeding habitat: Breeds in tundra and moorland in shallow peaty bog pools and sedge swamps with abundant bog-mosses.

LOOK-ALIKES (in range)
Hairy Hawker (*p. 136*)
Migrant Hawker (*p. 138*)
Blue-eyed Hawker (*p. 140*)
Moorland Hawker (*p. 148*)
Bog Hawker (*p. 150*)
Siberian Hawker (*p. 152*)
Baltic Hawker (*p. 154*)

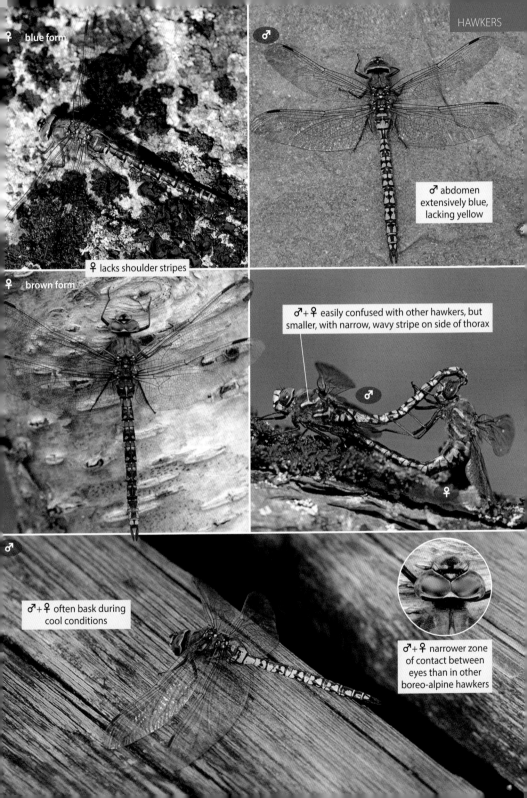

♀ blue form

♀ lacks shoulder stripes

♂

♂ abdomen extensively blue, lacking yellow

♀ brown form

♂+♀ easily confused with other hawkers, but smaller, with narrow, wavy stripe on side of thorax

♂

♀

♂+♀ often bask during cool conditions

♂+♀ narrower zone of contact between eyes than in other boreo-alpine hawkers

LC **Moorland Hawker**

Aeshna juncea ×1

Common Hawker

| Common (N), scarce (S) | = |

Lake, pond, bog

J F M A M J J A S O N D

| Overall length: | 65–80 mm |
| Hindwing: | 40–47 mm |

A powerful and aggressive 'mosaic hawker', the commonest and most widespread of a suite of large, boreo-alpine species. Although classified as 'stable' at the European scale, has declined significantly in Britain and the Netherlands in recent decades. Like other boreo-alpine species, it is threatened by climate change.

Identification: Broad yellow stripes on side of thorax; costa **bright yellow**; black line across 'face' **narrows** at eyes; thick stem to black 'T'-mark on top of frons; **yellow spot** behind eye (hard to see); broad contact between eyes (narrow on Azure Hawker (*p. 146*)). ♂ Rather dark, with pairs of bright blue dots and yellow flecks along abdomen; yellow shoulder stripes narrow but prominent; eyes bluish above, brown below. ♀ Eyes brown; shoulder stripes absent; pale markings on abdomen smaller than on male. **VARIATION:** Wings may show brownish suffusion with age; abdominal markings of female may be green or blue, yellow on immatures of both sexes.

Behaviour: Males are active even in overcast conditions, flying low over water and investigating edges of pools in search of females. Eggs are laid alone into aquatic vegetation, detritus or mud, typically above the waterline. Both sexes often feed high up in the open or along woodland rides well away from water, sometimes flying until dusk; rarely settles in the open.

Breeding habitat: Breeds mainly in acidic standing waters ranging from lakes to boggy pools on moorland and heathland, typically with abundant bog-mosses (*Sphagnum* spp.).

LOOK-ALIKES (in range)

Migrant Hawker (*p. 138*)
Blue Hawker (*p. 142*)
Azure Hawker (*p. 146*)
Bog Hawker (*p. 150*)
Siberian Hawker (*p. 152*)
Baltic Hawker (*p. 154*)

♂

♀

♂ + ♀ 'T'-mark on top of frons with thick stem

♂ + ♀ black line across 'face' narrows where meets eyes

♂ + ♀ costa yellow

♀ blue form

♀ i

♂ + ♀ diagnostic yellow spot behind eye (but hard to see)

LC **Bog Hawker**

Aeshna subarctica ×1

Subarctic Hawker

Local ▽

Bog, pond

J F M A M J J A S O N D

Overall length: 70–76mm
Hindwing: 39–46mm

Variant from C Europe showing additional yellow patches on side of thorax

A large 'mosaic' hawker almost identical to Moorland Hawker (*p. 148*), with which it co-exists in bogs, but appearing slightly darker. As with other boreo-alpine species, it is threatened by the warming and desiccating effects of climate change, especially in the S of its range.

Identification: Large hawker with black line across 'face' **as thick or wider** where meets eyes (narrows in Moorland Hawker); broad contact between eyes (narrow in Azure Hawker (*p. 146*)); thick stem to black 'T'-mark on top of frons; **lacks yellow spot** behind eye shown by Moorland Hawker; two broad yellowish stripes on side of thorax; pale abdominal markings smaller than on Moorland Hawker; costa **brownish**. ♂ Dark, with pairs of blue-green spots and yellow flecks along abdomen; yellowish shoulder stripes narrow but prominent; eyes bluish above, brown below. ♀ Brown, with similar abdominal pattern to male but all markings yellow; shoulder stripes very short or absent; eyes brown. VARIATION: In lowland bogs in C Europe, some individuals have additional **pale patches** either side of front stripe on side of thorax.

Behaviour: Males patrol territories over areas of waterlogged bog-mosses (*Sphagnum* spp.), where females lay eggs alone into floating mats.

Breeding habitat: Acidic bogs and pools with abundant bog-mosses, especially where it forms a floating 'soup'. The species occurs locally in NE Europe and more rarely farther S at lowland bogs and at high altitude (>700 m) in the Alps and W Carpathians; like many other species of NE Europe, its range extends across N Asia, and in this case also N America.

LOOK-ALIKES (in range)

Migrant Hawker (*p. 138*)
Azure Hawker (*p. 146*)
Moorland Hawker (*p. 148*)
Siberian Hawker (*p. 152*)
Baltic Hawker (*p. 154*)

♂

♂+♀ in C Europe, may have additional pale patches either side of front stripe on side of thorax, so thorax generally appears paler (as in image to *left*)

♂+♀ 'T'-mark on top of frons with thick stem

♂+♀ black line across 'face' as thick or wider where meets eyes

♀

♂

♀

♂+♀ costa brownish

NT Siberian Hawker

Aeshna crenata ×1

| Local | = |
| Pond, lake | |

J F M A M J J A S O N D

Overall length: 71–86 mm
Hindwing: 44–60 mm

The largest 'mosaic' hawker found only in a few localities in S Finland and the borders of Lithuania, Latvia, Belarus and Russia, although its range extends E right across Siberia. It is similar to Moorland Hawker (*p. 148*) in appearance, but males have all-blue markings on the abdomen.

Identification: Large hawker with thick stem to black 'T'-mark on top of frons; broad contact between eyes (narrow in Azure Hawker (*p. 146*)); **lacks yellow spot** behind eye shown by Moorland Hawker); two broad yellow stripes on side of thorax; costa **brownish**. ♂ Dark brown thorax; yellowish shoulder stripes narrow but prominent; black abdomen with pairs of **blue spots and flecks**; eyes bright blue-green; 2–3 large 'teeth' on top of straight upper appendages, near tip. ♀ Dark brown, with similar abdominal pattern to male but markings rather dull yellow-green; shoulder stripes short and obscure; eyes brown; appendages **pointed** (tips rounded in similar species); **dark suffusion** between node and wing-spot when mature. VARIATION: Females may occur in blue form.

Behaviour: Males patrol aggressively over lakeside territories. Unlike similar 'mosaic' hawkers, flight season is short, rarely extending beyond mid-September and peaking in late July–August.

Breeding habitat: Acidic ponds with fringing bog-mosses (*Sphagnum* spp.), and sedges (*Carex* spp.) within mature forest.

LOOK-ALIKES (in range)
Blue Hawker (*p. 142*)
Green Hawker (*p. 144*)
Azure Hawker (*p. 146*)
Moorland Hawker (*p. 148*)
Bog Hawker (*p. 150*)
Baltic Hawker (*p. 154*)

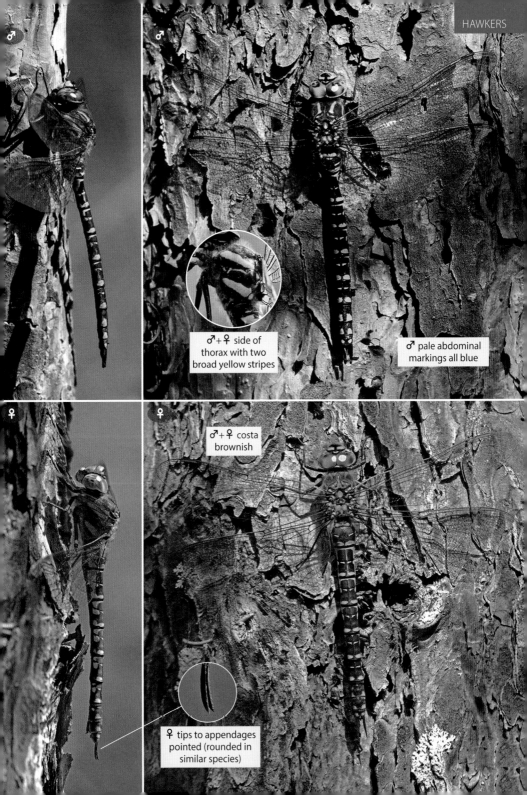

♂

♂

♂+♀ side of thorax with two broad yellow stripes

♂ pale abdominal markings all blue

♀

♂+♀ costa brownish

♀ tips to appendages pointed (rounded in similar species)

LC Baltic Hawker

Aeshna serrata ×1

Locally common	=
Lake, pond, river	

J F M A M J J A S O N D

Overall length:	75–81 mm
Hindwing:	48–51 mm

This large 'mosaic' hawker breeds in a restricted zone from Jutland, Denmark, eastwards, through inland waters in S Sweden to coastal Baltic wetlands.

Identification: Large hawker, almost as large as Siberian Hawker (*p. 152*), with **narrow** stem to black 'T'-mark on top of frons; black line across 'face' narrowing where meets eyes; broad contact between eyes (narrow in Azure Hawker (*p. 146*)); thorax has obvious yellow shoulder stripes and two broad yellow stripes on side; pale markings on abdomen larger and more rectangular than in similar species; costa **yellow**; cell at base of hindwing **evenly pale** (darker, two-toned in similar species). ♂ Appears bluer than Moorland Hawker, abdomen having larger blue markings, **rectangular** on S5–8, and small yellow flecks; from side, upper appendages upcurved with 4–7 small 'teeth' near tip (hence the scientific name '*serrata*'). ♀ Body dark brown with yellowish markings; **obvious** yellow shoulder stripes, unlike female Moorland Hawker. VARIATION: Female may have abdomen coloured blue and yellow, as male, and brownish suffusion in wings.

Behaviour: Slightly bulkier than Moorland Hawker, so flies more slowly. Tends to perch in dense vegetation around water bodies, rather than in trees.

Breeding habitat: Open, shallow, reedy ponds, lakes and slow-flowing rivers, including eutrophic and slightly brackish waters, and coastal reedbeds near the Baltic Sea.

LOOK-ALIKES (in range)

Blue Hawker (*p. 142*)
Green Hawker (*p. 144*)
Azure Hawker (*p. 146*)
Moorland Hawker (*p. 148*)
Bog Hawker (*p. 150*)
Siberian Hawker (*p. 152*)

cell at base of
hindwing evenly pale

♂ S5–8 with
large, rectangular
blue markings

♂

♂ + ♀ 'T'-mark on
top of frons with
narrow stem

♀ blue form has male-like
blue patterning on abdomen

♀ blue form

♀ shoulder stripes more obvious
than other female hawkers

♂ + ♀ costa yellow

♀

♀

LC Brown Hawker

Aeshna grandis ×1

Amber-winged Hawker

| Locally common | = |

Lake, pond, river, canal, ditch

J F M A M J J A S O N D

| Overall length: | 70–77 mm |
| Hindwing: | 41–49 mm |

A large, distinctive 'brown hawker' with golden-brown wings that are clearly visible in flight (see *below*). It breeds in mainly lowland standing waters, but in the S of its range occurs at altitudes up to 2,250 m. The peak flight period is in August, about two months later than the only other 'brown hawker', Green-eyed Hawker (*p. 158*).

Identification: Readily identified by mainly brown appearance, with **amber-tinted wings**, **golden veins** across front of wings and two **prominent** yellow stripes on side of thorax; both sexes lack pale shoulder stripes. ♂ Uniformly brown with small blue markings along side of abdomen and on S2; blue on top of brown eyes. ♀ Like male but with yellow markings along side of abdomen and yellowish-brown eyes. VARIATION: Stripes on side of thorax narrower in parts of Scandinavia; immatures as female; some females have blue markings on side of abdomen.

Behaviour: Patrols over water bodies and adjacent emergent vegetation but often found away from water, sometimes in large groups. In calm conditions, often flies at canopy level and is active late into evening. Flight action involves long glides interspersed with bursts of rapid, shallow wingbeats. Often settles briefly to eat prey. Unlike other hawkers, males rarely fight. Eggs are laid alone into either living or dead plant material, near or below water's surface. Emergence takes place at night and adults takes their first flight before dawn.

Breeding habitat: Variety of well-vegetated standing and slow-flowing water bodies, including ponds, lakes, gravel pits, canals and ditches, typically in woodland. Tolerates moderate levels of pollution.

LOOK-ALIKES (in range)
Green-eyed Hawker (*p. 158*)
Vagrant Emperor (*p. 172*)

♂

♂

♂ + ♀ side of thorax with two prominent yellow stripes

♀

♂ + ♀ wings have unmistakable amber suffusion

LC Green-eyed Hawker

Aeshna isoceles ×1

Norfolk Hawker

| Locally common | = |

| Lake, pond, canal, ditch |

J F M A M J J A S O N D

| **Overall length:** | 62–67 mm |
| **Hindwing:** | 39–45 mm |

A medium-sized 'brown hawker' that tends to fly earlier in the summer than the distinctly larger Brown Hawker (*p. 156*). In the west and north it is scarce or absent, although there are recent signs of range expansion in the N.

Identification: Large, ochre-brown hawker, only obvious markings on abdomen being narrow **yellow triangle** ('golf tee' shape) on S2 and narrow black line along remainder; sexes similar; eyes **green**; two yellow stripes on side of thorax; shoulder stripes yellow and small, absent on some; wings clear except for small orange-yellow area at base of hindwings and amber wing-spots and veins at front of wing. ♂ Rear edge of hindwing angular, like male hawkers (rounded in female). ♀ Stripes on side of thorax less prominent than those of male. VARIATION: Shoulder stripes bolder in SE Europe; stripes on side of thorax greenish on some.

Behaviour: Males patrol territories low along ditches or fringes of tall emergent vegetation, perching more often than other hawkers. Maturation takes up to three weeks, during which time immatures inhabit sheltered woodland edges and paths through tall vegetation. Females lay eggs alone, into living plants and usually just below the water's surface, rarely into dead plant material.

Breeding habitat: Found in swamps, ditches, canals, ponds and lakes with beds of Common Reed (*Phragmites australis*), bulrushes (*Typha* spp.), sedges (*Carex* spp.) and (in N) dense Water-soldier (*Stratiotes aloides*). Occasional at slow-flowing water.

LOOK-ALIKES (in range)
Brown Hawker (*p. 156*)
Vagrant Emperor (*p. 172*)

♂ + ♀ yellow triangle on S2

♂ + ♀ green eyes (often obvious in flight)

A 'brown hawker' with clear wings flying in late spring/ early summer is most likely to be this species.

♂ + ♀ stripes on side of thorax green on some

NT Eastern Spectre

Caliaeschna microstigma ×1

Scarce and local ▽

Stream, river

J F M A M J J A S O N D

Overall length:	50–60 mm
Hindwing:	35–41 mm

A small, well-patterned dragonfly found at shady, flowing waters in SE Europe – it is the only hawker-like species to be found in such habitat (apart from Cretan Spectre (*p. 164*) in Crete). Although in a different genus to the other spectres, like them it has a 'tail light' across the tip of the abdomen and is rarely seen in sunlight.

Identification: Slightly smaller than Hairy Hawker (*p. 136*) and with similar pattern; two wide yellow stripes on side of thorax and pale shoulder stripes; abdomen has small, paired (blue or yellow) markings, larger on S9–10, creating a **'tail light'**; wing-spots **notably small** and dark. ♂ Shoulder stripes **blue and sharply hooked** at rear; abdomen black with **all-blue** markings, mostly small but increasingly large on S8–10, **merging on S10**; eyes blue. ♀ Dark brown with yellow shoulder stripes and yellowish markings on abdomen, those on **S9–10 wholly pale**; eyes greenish; appendages **tiny**; ovipositor sheath reaches beyond S10. VARIATION: Some females have pale blue markings on the abdomen; immatures similar in appearance to female.

Behaviour: Males patrol very low along the edges of suitable shady waters, like other spectres. They fly rather slowly but erratically, looking carefully for egg-laying females. Especially active during second half of the day, even at dusk. Mating takes place in trees.

Breeding habitat: Wooded streams and small rivers that flow all year round. These are often in hilly regions and are typically shaded, stony and fast-flowing with plenty of exposed roots.

LOOK-ALIKES (in range)

Hairy Hawker (*p. 136*)
Migrant Hawker (*p. 138*)
Blue-eyed Hawker (*p. 140*)

♂ ♂ pale shoulder stripes sharply hooked

♂ + ♀ wing-spots small and dark

♂ blue abdominal patterning with 'tail light' distinctive

♂i

♀ ♀ appendages very short

♀ S9–10 all-pale (obvious 'tail light')

LC Western Spectre

Boyeria irene ×1

Dusk Hawker

| Locally common = |

| River, stream, lake |

J F M A M J J A S O N D

| **Overall length:** | 63–71 mm |
| **Hindwing:** | 39–45 mm |

A fairly large but dull hawker-like spectre with subtle markings. Hard to spot due its cryptic patterning, unobtrusive behaviour and distinct preference for shade.

Identification: Body brown with pale markings, including shoulder stripes, two pale stripes on side of thorax and blotches on abdomen, some forming irregular bands. ♂ Pale markings bluish-green, including **'tail light'** on S9–10; wingtips **dusky**; eyes green. ♀ Pale markings yellowish; appendages **usually very short** (length of S10); eyes dull greenish. VARIATION: Some females have dusky wingtips and some have very long appendages (3× length of S10); immatures have pale yellowish patterning.

Behaviour: Males patrol tirelessly in search of females, flying low, close to water margins and often in the shade of overhanging trees. Eggs are laid into moss, *etc.* on rocks and tree roots. Perches in shade on trees or rock-faces. Most active late in the day, gathering in clearings to feed in groups, some flying after dusk and occasionally attracted to lights.

Breeding habitat: Streams and rivers usually with well-wooded banks; in the Alps, only found in large lakes with wave-splashed margins.

LOOK-ALIKES (in range)
Hairy Hawker (*p. 136*)
Migrant Hawker (*p. 138*)
Blue-eyed Hawker (*p. 140*)
Blue Hawker (*p. 142*)

♂ wingtips dusky

♂ S9–10 'tail light' more-or-less obvious

♀ appendages typically short but can be very long in some (see *below*)

Spectres compared *pp. 129, 131, 134*

E EN **Cretan Spectre**

Boyeria cretensis × ?

Scarce, very local ▽

Stream

J F M A M J J A S O N D

Overall length:	69–71 mm
Hindwing:	44–47 mm

Endemic to Crete. Almost identical to Western Spectre (*p. 162*), being a fairly large but cryptically patterned, hawker-like spectre with a similar preference for shade, but the ranges of the two species do not overlap.

Identification: Body brown with pale markings, including shoulder stripes, two pale stripes on side of thorax and blotches on abdomen. Very slightly larger and darker overall than Western Spectre, with smaller pale markings (especially bases of S2–8) and even more mottled appearance; angle at top of frons more **pointed**. ♂ Pale markings dull yellow-green, including '**tail light**' on S9–10; wingtips **dusky**; eyes green. ♀ Pale markings yellowish; appendages short (length of S10); eyes dull greenish. VARIATION: Immatures have more extensive pale yellowish patterning on abdomen.

Behaviour: Males patrol the margins of shady sections of streams tirelessly in search of females. Eggs are laid into mosses and tree roots. Most often seen during late afternoon, evening and at dusk; may be attracted to lights.

Breeding habitat: Rocky streams with fast-flowing water and at least partly shaded banks.

LOOK-ALIKES (in range)
Migrant Hawker (*p. 138*)
Blue-eyed Hawker (*p. 140*)

♂ smaller coloured patches on abdomen than Western Spectre, but 'tail light' still evident

CRETAN SPECTRE

WESTERN SPECTRE

♂+♀ top of frons more pointed than on Western Spectre

LC Blue Emperor

Anax imperator × 1

Emperor Dragonfly

Very common △

Lake, pond, river, canal

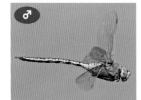

J F M A M J J A S O N D

Overall length:	66–84 mm
Hindwing:	45–51 mm

One of the largest and most colourful dragonflies, this aggressive species emerges in late spring and flies for much of the summer in seemingly unending patrols well above the water's surface. Its range has expanded significantly to the N in recent decades. Across most of Europe, territorial brightly coloured large dragonflies flying before July are most likely to be this species or Lesser Emperor (*p. 170*).

Identification: Both sexes: **apple-green** thorax **without** obvious black markings; **yellow** costa; brown wing-spots; often fly with **drooping** abdomen, which has irregular thick black **line** along top of S2–10; **pentagonal** black marking on top of 'forehead'. ♂ Abdomen has **green** S1, **otherwise bright blue**; blue 'wedge' shapes in front of forewing bases; eyes blue-green; wings clear. ♀ Abdomen mainly dull green; eyes greenish. VARIATION: Blue of male abdomen fades in cool conditions; female abdomen often blue, as male, and wings become brownish with age; immatures pale green with brown on abdomen.

Behaviour: Territorial males patrol waters for long periods, typically flying far from the shore and higher than 'mosaic' hawkers, and vigorously chasing away other emperors and hawkers. Most prey eaten in flight. Females most often seen when egg-laying alone (unlike other emperors), often well away from the margins, typically into floating vegetation and pondweed just below the water's surface.

Breeding habitat: A variety of well-vegetated standing and slow-flowing rivers in lowlands. Will breed in new ponds and can tolerate brackish water.

LOOK-ALIKES (in range)
Lesser Emperor (*p. 170*)
Vagrant Emperor (*p. 172*)
'Mosaic' hawkers (*pp. 136–155*),
  esp. Green Hawker (*p. 144*)
Common Green Darner (*p. 168*)

♂ + ♀ abdomen with dark line along top from S2–10

Only likely to be confused with the very rare vagrant Common Green Darner (see *p. 169* for comparison).

♂ + ♀ thorax uniformly apple-green

♂ + ♀ costa yellow

♀ blue form

♂

♀

♀

Common Green Darner

Anax junius ×

Vagrant
Lake, pond

J F M A M J J A S O N D

Overall length:	68–80 mm
Hindwing:	45–56 mm

♂

The commonest N American dragonfly and a migratory species prone to displacement by autumn storms. A few have successfully crossed the Atlantic to reach European shores. The first European record involved at least four males and four females in SW England in September 1998, associated with a hurricane; a male was recorded on the French coast near Nantes in September 2003, and a female was found on Corvo, Azores, in October 2019.

Identification: Very like Blue Emperor (*p. 166*), but both sexes have **brownish** eyes, green thorax and distinctive black and yellow '**bull's-eye**' pattern on top of 'forehead' (black pentagonal marking on Blue Emperor). Like Lesser Emperor (*p. 170*), abdomen **brightest at base**, colour fading quickly towards the tip and with gradually widening dark pattern; abdomen becomes purplish in cool conditions. ♂ Thorax in front of forewing bases lacks blue 'wedges' of Blue Emperor; abdomen has pale centre to S2 **without a black central line**; upper appendages long (as long as S9+S10), with **tiny spine** at tip of each; lower appendage very short (<20% of uppers). ♀ Side of abdomen brownish or blue-grey, base green; two tiny lumps (tubercles) on rear of head (absent in Blue Emperor but present in Lesser Emperor). VARIATION: Immatures have red-brown or violet abdomen.

Behaviour: In N America on the wing spring–autumn in N, all year in S. Many migrate, moving N to breed in early spring, next generation returning S in immature colours in autumn. They migrate during the day, covering about 10 km per day. Pairs egg-lay in tandem (unlike Blue Emperor).

Breeding habitat: Variety of standing waters and slow-flowing rivers, including brackish and temporary waters; vagrants anywhere sheltered from storms.

LOOK-ALIKES (in range)
Blue Emperor (*p. 166*)
Lesser Emperor (*p. 170*)

♂

♂

♂+♀ abdomen with dark line along top from S3–10, tapering towards the front

♂ S2 lacks dark line along top (unlike Blue Emperor (*p. 166*))

COMMON GREEN DARNER	♂+♀ round 'bull's-eye' pattern in front of the eyes		♂ appendages with tiny spine at tip		♀ two small lumps (tubercles) on rear of head	
BLUE EMPEROR	♂+♀ pentagonal black marking in front of the eyes		♂ appendages blunt-tipped		♀ no small lumps (tubercles) on rear of head	

♀

♀

LC **Lesser Emperor**

Anax parthenope ×1

Yellow-winged Emperor

Locally common △

Lake, pond, canal

J F M A M J J A S O N D

Overall length:	62–75 mm
Hindwing:	44–51 mm

The overall brown appearance of this large dragonfly is relieved by a bright blue 'saddle' at the base of the abdomen. Its range has expanded northwards somewhat in recent decades, although it is still scarcer in the N than farther S.

Identification: Slightly smaller than Blue Emperor (*p. 166*), with **green** eyes and plain brownish thorax; abdomen, typically held **straight**, has irregular thick **black line** along top and narrow **yellow band** at front of S2; wings have **yellowish suffusion** on outer half, **yellow** costa and brown wing-spots. ♂ **Pale blue 'saddle'** on S2 extending down sides and on to S3; black line along top of abdomen ends in point on waisted S3; sides of S4–10 olive-brown or dull blue-green. ♀ Abdomen usually duller than male with black central marking extending onto S2; two diagnostic tiny black **protrusions** on back of head, between the eyes (hard to see, even in photographs). VARIATION: Side of thorax may be greenish; yellow ring on S2 may be missing; females may have blue base and side of abdomen; immatures have brownish eyes and green abdominal markings.

Behaviour: Flight less powerful than Blue Emperor, which dominates in territorial clashes. Lays eggs in tandem into floating plants and debris.

Breeding habitat: Typically breeds in lakes, ponds and slow-flowing waters; tolerates brackish water.

LOOK-ALIKES (in range)
Other emperors (*pp. 177–173*)

♂i

♂i

♂ S2 blue with yellow band across front

♂+♀ eyes green

♂+♀ wings with yellowish suffusion on outer half

♀

♀ blue form

♀ black central marking on abdomen extends onto S2

Emperors flying or egg-laying in tandem will either be this species or (in S) Vagrant Emperor.

♂+♀ costa yellow

♂

♀

♂+♀ thorax brownish

LC Vagrant Emperor

Anax ephippiger ×1

Scarce and erratic △

Lake, pond

J F M A M J J A S O N D

Overall length:	61–70mm
Hindwing:	43–48mm

Although it breeds sporadically in S Europe, occurrences farther N of this wanderer from arid lands are generally associated with dust-laden winds from the Sahara, even during the winter. This is the only dragonfly known to have reached Iceland.

Identification: Slightly smaller than similar Lesser Emperor (*p. 170*), but largely **yellowish-brown** with **brown** eyes; thorax brown or greenish; abdomen relatively short and slender, with irregular black marking along top from S3–10 and pairs of **yellowish spots** on S8–10; wing-spots long and orange; costa **yellow**; hindwings broad, variably **suffused** with yellow; upper appendages have pointed tips, broader on females. ♂ S2 **bright blue on top**, producing more restricted 'saddle' than on Lesser Emperor. ♀ Darker and duller than male; black marking on top of abdomen continues onto S2, which is **no more than tinged violet**. VARIATION: May have yellowish ring on base of S2; immatures may have yellow-green thorax and obscure 'saddle' on S2.

Behaviour: A nomadic species and an opportunistic breeder whose larvae are able to develop rapidly, enabling it to take advantage of transient wet conditions, particularly in hot climates. Eggs are usually laid in tandem, sometimes outside normal range, but larvae rarely survive the winter, even in S. May be seen in all months around the Mediterranean, but farther N most records are associated with strong S airflows originating in North Africa. Disperses long distances and may be attracted to lights, including moth traps, when migrating at night.

Breeding habitat: Ponds, lakes and marshes, often shallow, seasonally wet and brackish. Breeds sporadically in S Europe and occasionally as far N as S England following widespread invasions.

LOOK-ALIKES (in range)

Lesser Emperor (*p. 170*)
Brown Hawker (*p. 156*)
Green-eyed Hawker (*p. 158*)

♂ + ♀ eyes brown

♂ S2 blue; rest of abdomen brownish with black central line

♂ + ♀ costa yellow

♂ + ♀ wings variably suffused yellow

♂ + ♀ thorax brown or greenish

♂ + ♀ yellowish spots on sides of S8–10

Emperors compared *pp. 129, 131, 135*

VU Magnificent Emperor

Anax immaculifrons ×1

Fiery Emperor

Rare and local ?

Stream, pond

J F M A M J J A S O N D

Overall length:	80–86 mm
Hindwing:	54–60 mm

♂

This aptly named dragonfly is spectacular both in size and colour as it patrols rocky streams. It is known only from the Greek Islands of Rhodes, Karpathos and Ikaria, Cyprus and the S coast of Turkey, although also occurs widely across S Asia.

Identification: Very large with bold body pattern; **three black stripes** across side of otherwise pale thorax; pale bases to S2–8 creating **seven bold 'rings'** on abdomen; yellow costa and many small veins in forewings. ♂ Eyes **bright blue**; frons ('face') and top of thorax pale blue-green; thorax side and S1–2 **pale blue**; rest of abdomen black with **pinkish-yellow** covering 50–70% of bases of S3–8; wings tinged golden, enhanced by extensive yellow venation. ♀ Eyes greenish on top; body patterned much as male but **wholly black-and-yellow**. VARIATION: Female may have some yellow suffusion in wings.

Behaviour: Males patrol fast over long territories, pausing at and circling ponded areas. Uniquely for a hawker or emperor, males hover over females to guard them while they lay eggs. Perches on shady rock-faces.

Breeding habitat: Small but permanent rocky streams in hilly areas, including deeper and ponded sections with relatively still water.

LOOK-ALIKES (in range)
Blue-eyed Goldenring (*p. 224*)

♂ + ♀ side of thorax has three black stripes

♂ eyes blue

♂ appears bluish at the front, black and pinkish-yellow at the rear

♀ boldly patterned in black and yellow

CLUBTAILS, PINCERTAILS, SNAKETAIL, HOOKTAIL & BLADETAIL

FAMILY | **Gomphidae**

GENERA | *Gomphus, Stylurus, Onychogomphus, Ophiogomphus, Paragomphus & Lindenia*

15 SPECIES

Mostly medium-sized 'perchers' (mean lengths 44–75 mm), with eyes clearly separated and complex black-and-yellow patterning. Abdomen tip typically swollen, some with prominent, curved male appendages; auricles present on sides of S2 of males. Rear margin of the male's hindwing is acutely angled (as in 'mosaic' hawkers (*pp. 136–155*)). Most species breed in running water, but are frequently found well away from breeding waters. Females lay eggs by repeatedly dipping abdomen tip into water, unaccompanied by males. The five 'types' (from six genera) can be differentiated by careful examination of patterning (particularly the black patterning on the thorax) and abdomen shape, especially male appendages. A good photograph is useful to check details that may not be visible in the field.

Key features of Gomphidae

- ◆ eyes widely separated
- ◆ mostly medium-sized 'perchers'
- ◆ complex black-and-yellow patterning
- ◆ prominent curved male appendages (most species)
- ◆ breed mostly in flowing water
- ◆ often found away from water

Identify 'sub-types' by shape of male appendages | pattern and shape of abdomen | pattern on top and side of thorax

Clubtails

7 species

GENERA: *Gomphus (pp. 178–187)* & *Stylurus (pp. 188–191)*

KEY FEATURES:

- ◆ thorax yellow or pale green
- ◆ abdomen with narrow yellow markings on top
- ◆ male abdomen often swollen at tip
- ◆ male upper appendages diverge

Identify species by patterning on thorax and abdomen

Clubtail

Clubtails (thorax sides) compared

narrow	narrow	very narrow
Common Clubtail *Gomphus vulgatissimus* (p. 178)	**Turkish Clubtail** *Gomphus schneiderii* (p. 180)	**Pronged Clubtail** *Gomphus graslinii* (p. 184)

thin black lines / wavy, continuous	long	lines join / broad	lines join / broad
Western Clubtail *Gomphus pulchellus* (p. 186)	**Yellow Clubtail** *Gomphus simillimus* (p. 182)	**River Clubtail** *Stylurus flavipes* (p. 188)	**Syrian Clubtail** *Stylurus ubadschii* (p. 190)

Pincertails

5 species

GENUS: *Onychogomphus (pp. 194–203)*

KEY FEATURES:

◆ thorax yellow or buff
◆ abdomen with broad yellow markings on top
◆ male abdomen slightly swollen at tip
◆ male upper appendages curved, converging at tip

Identify species by patterning on thorax and abdomen | shape of male appendages

Pincertail

Pincertails (thorax and male appendages from side) compared

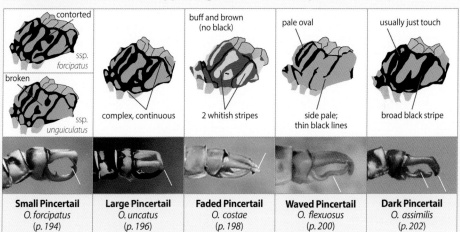

contorted — ssp. *forcipatus*
broken — ssp. *unguiculatus*

buff and brown (no black) | 2 whitish stripes

complex, continuous

pale oval | side pale; thin black lines

usually just touch | broad black stripe

Small Pincertail	**Large Pincertail**	**Faded Pincertail**	**Waved Pincertail**	**Dark Pincertail**
O. forcipatus	*O. uncatus*	*O. costae*	*O. flexuosus*	*O. assimilis*
(p. 194)	(p. 196)	(p. 198)	(p. 200)	(p. 202)

Snaketail

1 species

GENUS: *Ophiogomphus (p. 192)*

KEY FEATURES:

◆ face, eyes and thorax green when mature
◆ abdomen with tapering yellow markings and bulging tip
◆ male appendages parallel when seen from above

Snaketail

Hooktail

1 species

GENUS: *Paragomphus (p. 204)*

KEY FEATURES:

◆ thorax green when mature
◆ abdomen with indistinct dark markings on brown, and large 'flaps' on sides of S8–9
◆ male upper appendages long and downcurved

Hooktail

Bladetail

1 species

GENUS: *Lindenia (p. 206)*

KEY FEATURES:

◆ abdomen pale yellow, darkening and becoming pruinose with age
◆ abdomen with large 'flap' on sides of S7–8
◆ male upper appendages long, straight and parallel

Bladetail

177

LC Common Clubtail

Gomphus vulgatissimus ×1‹

Club-tailed Dragonfly

Common	=
River, stream, lake	

J F M A M J J A S O N D

Overall length:	45–50mm
Hindwing:	28–33mm

♂ + ♀ thorax: pale shoulder stripe narrower than black line either side

A medium-sized dragonfly with black-and-yellow or black-and-green patterning, found in early summer at, or in the vicinity of, silty sections of rivers. Like other clubtails, often found well away from water. The most widespread clubtail, although absent from the far NW and S of Europe.

Identification: Medium-sized with **greenish** eyes and bold black-and-yellow or black-and-lime green patterning; **narrow** pale shoulder stripe between broad black stripes; discontinuous, mainly narrow yellow markings along top of S1–7, **top of S8–10 all-black**; legs black; darker overall than other clubtails. ♂ Black and **pale green** when mature, except for yellow spots on sides of S8–9; abdomen is waisted at S3 and distinctly clubbed at S8–9. ♀ Abdomen thicker and less obviously clubbed than in male, with broader yellow markings. VARIATION: Black shoulder stripes may touch.

Behaviour: Adults most easily found during synchronous emergence in spring, before dispersing up to 10 km to mature away from water. They often perch in trees, where difficult to observe. Territorial males fly fast and low over water, returning to sit on exposed perches.

Breeding habitat: Breeds in slow-flowing stretches of rivers and streams where silt accumulates. Can be locally common in large, well-oxygenated standing waters but avoids fast-flowing, rocky watercourses.

LOOK-ALIKES (in range)
Other clubtails (*pp. 180–191*)
Green Snaketail (*p. 192*)
Pincertails (*pp. 194–203*)
Orange-spotted Emerald (*p. 248*)
♀ skimmers (*pp. 262–281*)

♂

♂+♀ top of S8–10 black   ♂+♀ eyes greenish

♀

♀   ♂+♀ pale shoulder stripe narrow (see *opposite*)   ♂

NT Turkish Clubtail

Gomphus schneiderii ×1

Scarce and local ?

River, stream, lake

J F M A M J J A S O N D

Overall length:	40–48 mm
Hindwing:	29–31 mm

♂ + ♀ thorax: pale shoulder stripe almost as wide as black line either side

A medium-sized dragonfly with black-and-yellow or black-and-green patterning, found in far SE of Europe, overlapping with very similar Common Clubtail (*p. 178*) in the Balkan Peninsula, where intermediate forms occur and identification may be impossible. Future studies may conclude that it is a subspecies of Common Clubtail.

Identification: Medium-sized with **blue** eyes and bold black-and-yellow or black-and-lime green patterning; pale shoulder stripe **almost as wide as** adjacent black stripes; discontinuous, mainly narrow yellowish markings along top of **S1–9**, but very small or absent on S8–9; legs black. ♂ Black-and-yellow or black-and-pale green; abdomen waisted at S3 and distinctly clubbed at S8–9. ♀ Abdomen thicker and less obviously clubbed than in male, with broader yellow markings than Common Clubtail. VARIATION: Females may have some yellow on legs.

Behaviour: As Common Clubtail, often perches in trees, where may be difficult to see. Territorial males often sit on tips of branches overhanging water and other exposed perches.

Breeding habitat: Breeds in slow-flowing rivers and streams where silt accumulates, perhaps in warmer conditions than Common Clubtail. Present in some large, well-oxygenated standing waters.

LOOK-ALIKES (in range)
Other clubtails (*pp. 178–191*)
Pincertails (*pp. 194–203*)
Slender Skimmer (*p. 280*)

♂ + ♀ top of S8–9
may have small
yellow markings

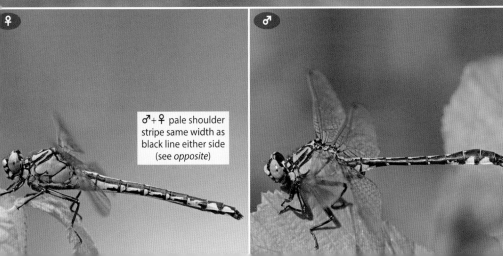

♂ + ♀ pale shoulder
stripe same width as
black line either side
(see *opposite*)

NT Yellow Clubtail

Gomphus simillimus ×1

Locally common ▽

River, stream, canal

J F M A M J J A S O N D

Overall length:	45–50 mm
Hindwing:	29–33 mm

♂ + ♀ thorax: pale shoulder
stripe same width as black line
either side

This rather bright clubtail is found to the south and west of NE France and the upper Rhine Valley. It is most common in SW France, where it can be found at large rivers and other slow-flowing waters.

Identification: Similar to other clubtails in overall black-and-yellow pattern. Eyes **blue**; pale shoulder stripe **as wide as** adjoining black lines and linked to yellow on top of thorax (but **not** extending down as far as legs), yellow 'collar' interrupted by thin black line and narrow black line on side **forking** before reaching base of hind legs; abdomen has interrupted yellow markings on top of **all** segments; costa yellow; legs with **some yellow** (as River Clubtail (*p. 188*)). ♂ Abdomen and **yellow markings swollen at S8–9** (as Common Clubtail (*p. 178*) and River Clubtail). ♀ Abdomen thicker and less obviously clubbed than in male, with broader yellow markings than Common Clubtail; vulvar scale longer than in other clubtails, about **half length of S9**.

Behaviour: Males perch on rocks or vegetation beside water. Has wandered far from breeding areas, including into Belgium (where not thought to be breeding).

Breeding habitat: Mainly slow-flowing waters in lowlands, especially large rivers. Also found at streams, canals and sometimes lakes and ponds.

LOOK-ALIKES (in range)
Other clubtails (*pp. 178–191*)
Pincertails (*pp. 194–203*)
Orange-spotted Emerald (*p. 248*)
♀ skimmers (*pp. 262–281*)

♂

♂ + ♀ pale shoulder stripe as wide as black line either side (see *opposite*)

♂ + ♀ eyes blue

♀

♂ + ♀ yellow on all abdominal segments

♂

♀

♀ vulvar scale half length of S9 (much longer than in other clubtails)

E **NT** **Pronged Clubtail**

Gomphus graslinii ×1

PROTECTED

Scarce and local ▽

River

J F M A M J J A S O N D

Overall length:	47–51 mm
Hindwing:	27–30 mm

♂ + ♀ thorax: pale shoulder stripe much narrower than black line either side

This W European endemic clubtail is restricted to SW France and, more sparingly, in the Iberian Peninsula. The strongest populations are in the catchments of the Hérault, Lot and Tarn.

Identification: Similar to other clubtails in overall black-and-yellow pattern, but most like Yellow Clubtail (*p. 182*). Eyes **blue**; pale shoulder stripe **narrower than** adjoining black lines, upper of which is very broad and usually joins black line at top of thorax (as River Clubtail (*p. 188*)); legs with **yellow only on femora** (not tibiae). ♂ Abdomen with **S8–9 only slightly 'clubbed'** and yellow **wine-goblet shape on S9**; upper appendages with short **'prong' on outer edge**. ♀ Marking on S9 broader than in male, but still with yellow across tip of segment. VARIATION: Males may become greenish with age.

Behaviour: Flies later in summer than other clubtails in its range and, like others, can often be most easily found away from water in nearby bushes and tall vegetation.

Breeding habitat: Mainly slow-flowing sections of large rivers and streams with deposits of silt and organic detritus and wooded or scrubby margins; sometimes in large drains and reservoirs, and often in hilly country. In recent decades, has been lost from N parts of range, but many new sites have been discovered in the Iberian Peninsula. Threatened by increasing frequency of droughts due to climate change.

LOOK-ALIKES (in range)

Other clubtails (*pp. 178–191*)
Pincertails (*pp. 194–203*)
Orange-spotted Emerald (*p. 248*)
♀ skimmers (*pp. 262–281*)

♂ 'prong' on sides of upper appendages

♂ S8–9 only slightly 'clubbed'; 'wine-goblet' shape on S9

♂ + ♀ eyes blue

♀ broad wine-goblet shape on S9, including yellow across tip of segment

♂ + ♀ pale shoulder stripe very narrow (see *opposite*)

♂ + ♀ legs with yellow on femora only (not on tibiae)

LC Western Clubtail

Gomphus pulchellus × 1·

Common =

Lake, river, canal

J F M A M J J A S O N D

| Overall length: | 47–50 mm |
| Hindwing: | 27–31 mm |

A rather dull but distinctive clubtail which is unusual amongst the clubtails in that it is often found at standing waters. It is endemic to W Europe and has spread NE to Austria, Germany and the Netherlands in the last century, probably aided by the creation of gravel pits and canals, as well as climate change.

Identification: ♂ & ♀ pale yellow-and-black, markings similar; thorax has **fine** black lines on shoulder and side (similar to Green Snaketail (*p. 192*)), with **wavy** line between forewing base and middle leg; abdomen with **almost parallel sides**, lacking obvious clubbed tip of other clubtails, and with discontinuous **yellow stripe along top of all segments**; costa yellow; legs **yellow**-and-black; eyes pale blue-green. VARIATION: Some males greenish-yellow.

Behaviour: Males patrol shoreline territories, flying quite fast, low and erratically over water with bouncing action; rarely hover. Adults frequently settle on bare ground or a low perch a short distance from the water, where subdued colours make them difficult to see.

Breeding habitat: Prefers sluggish rivers and associated standing waters such as oxbow lakes. In N of range, more often in standing waters such as fishponds, flooded mineral workings and canals.

♂ + ♀ thorax: fine black lines on shoulder and side; wavy line from base of forewing to middle leg

LOOK-ALIKES (in range)

Other clubtails (*pp. 178–191*)
Orange-spotted Emerald (*p. 248*)
♀ skimmers (*pp. 262–281*)

186

♂ abdomen almost parallel-sided
and yellow coloration rather faded

♂+♀ abdomen with yellow
stripe along top of all segments

♂+♀ side of thorax has wavy
black line across side (see *opposite*)

♂+♀ legs yellow-and-black

LC River Clubtail

Stylurus flavipes ×1∙

Yellow-legged Clubtail

PROTECTED
Locally common △
River, canal, drain

J F M A M J J A S O N D

Overall length:	50–55 mm
Hindwing:	30–35 mm

♂ + ♀ thorax: pale shoulder stripe much broader than black line either side; upper black line joined to black line on top; line from hindwing base unforked

The largest clubtail and the hardest to find. It occurs at large rivers from France eastwards into N Asia, becoming commoner in E Europe. It has recolonized many rivers in W and C Europe since the 1990s, possibly due to improved water quality and river management, and climate change.

Identification: Medium-sized with bold black-and-yellow patterning and extensive areas of yellow on legs. Abdomen yellower, longer and slimmer than Common Clubtail (*p. 178*). ♂ & ♀ Eyes blue in mature male, green in female; thorax has six black stripes on top, central pair separated by central yellow line that joins **yellow collar** at front, forming 'T'-shape when seen from above; central black stripes **join** next ones down at both ends, enclosing two long **yellow ovals**; yellow shoulder stripe **broader** than bordering black lines; side of thorax has thin, **unforked**, black line from hindwing base to just rear of hind leg; abdomen has broader yellow markings on top than Common Clubtail, extending discontinuously to clubbed **S8 & S9** (black in Common Clubtail); costa dark. VARIATION: Eyes brown in immatures.

Behaviour: Emerges in midsummer, flying later than most other clubtails, especially in N. Adults stay close to breeding waters, in bushes and rough vegetation. Males patrol territories low over water, well away from margins, but may sit on sandbanks.

Breeding habitat: Larvae live in sand in sluggish sections of large rivers, as far downstream as their tidal limits. May also use canals and drains with similar sandy sediment.

LOOK-ALIKES (in range)
Other clubtails (*pp. 178–191*)
Green Snaketail (*p. 192*)
♀ pincertails (*pp. 194–203*)
Orange-spotted Emerald (*p. 248*)
♀ skimmers (*pp. 262–281*)

♂+♀ yellow 'T' shape and two yellow ovals on top of thorax (also see *opposite*)

♂+♀ top of S8–9 yellow

♂+♀ long, slender abdomen

♂+♀ legs yellow-and-black

Syrian Clubtail

Stylurus ubadschii ×1·

| Rare ? |
| River |

J F M A M J J A S O N D

| Overall length: | 41–55 mm |
| Hindwing: | 25–31 mm |

A large clubtail that replaces the very similar River Clubtail (*p. 188*) to the south of that species' range, occurring from SW Turkey eastwards. Like River Clubtail, it is placed within a genus found principally in N America.

Identification: Medium-sized with bold black-and-yellow patterning; **eyes blue**; yellow 'T'-shape and two **ovals** on top of thorax; yellow shoulder stripe **broader** than bordering black lines; side of thorax has thin, **unforked** black line from hindwing base to just rear of hind leg; long abdomen with broader yellow markings on S8 & S9 than co-occurring Turkish Clubtail (*p. 180*); legs with **less yellow** than River Clubtail. ♂ Abdomen slimmer than in River Clubtail, with **more obvious 'club'** extending from S7–9 (S8–9 in River Clubtail). ♀ Eyes blue (green in River Clubtail).

Behaviour: Adults forage around riverside shrubs. Females lay eggs over sandy areas in the riverbed.

Breeding habitat: Small numbers have been recorded since at least 2012 (perhaps since 2000) from the Eşen Çayi, SE Muğla province, Turkey. Found in large rivers, much as River Clubtail, but may wander away from water, as other clubtails.

♂ + ♀ thorax: pale shoulder stripe much broader than black line either side; upper black line joined to black line on top; line from hindwing base unforked

♂

♂ S7–9 conspicuously swollen, top of S8–9 yellow

♂ + ♀ yellow 'T' shape and two yellow ovals on top of thorax (also see *opposite*)

♀

♂

♂ + ♀ eyes blue

♀

♂ + ♀ yellow on legs more restricted than in River Clubtail

LC Green Snaketail

Ophiogomphus cecilia ×1

Green Clubtail

PROTECTED

Locally common =

River

J F M A M J J A S O N D

Overall length:	50–60 mm
Hindwing:	30–36 mm

The only snaketail in Europe; resembles a large clubtail in general coloration and river habitat, but the 'front-end' is vivid apple-green in mature individuals and the male appendages are short. Found mainly in E Europe, becoming more localized in C and SE Europe.

Identification: Largest of the Gomphidae apart from Bladetail (*p. 206*), with **green** eyes and 'face', and **green** thorax with thin black lines on top and side, the latter similar to those on Yellow Clubtail (*p. 182*) and Western Clubtail (*p. 186*); base of abdomen **green extending to S2**, rest of abdomen has discontinuous yellow pattern to S10, generally broader than on clubtails (*pp. 178–191*) and those on S3–7 more triangular in shape, and moderately clubbed at S8–9; legs extensively **yellow**. ♂ Appendages **short and yellowish**. ♀ Yellow markings on abdomen broader than on male; two tiny '**crests**' on back of head, between eyes. VARIATION: Black lines on thorax reduced in some; immatures lack any green.

Behaviour: Males patrol over large rivers or perch beside smaller watercourses.

Breeding habitat: Mainly meandering rivers in the lowlands with natural flows, few or no aquatic plants and, most importantly, deposits of sand and gravel in which the larvae live; rarely found in canals, and avoids waters with fast currents or fine sediment, favouring stretches where the banks are at least partly unshaded.

LOOK-ALIKES (in range)
Clubtails (*pp. 178–191*)
♀ Small Pincertail (*p. 194*)
♀ skimmers (*pp. 262–281*)

♂ + ♀ vivid green 'front-end'
very distinctive in when mature

♂ + ♀ thin lines on thorax
similar to Yellow Clubtail (*p. 182*)
and Western Clubtail (*p. 186*)

LC Small Pincertail

Onychogomphus forcipatus ×1

Green-eyed Hooktail, Green-eyed Hook-tailed Dragonfly

Common =
(ssp. *albotibialis* ▽)

River, stream

J F M A M J J A S O N D

Overall length:	46–50 mm
Hindwing:	25–30 mm

ssp.
forcipatus

ssp.
unguiculatus

♂+♀ thorax: contorted black
line across centre of side, often
broken

The most widespread pincertail, and the only one that occurs over much of E and C Europe. Confusingly variable: black markings are extensive in N of its range (subspecies *forcipatus*), where it is less common, but more restricted in southern populations (subspecies *unguiculatus* in SW Europe and *albotibialis* (NT) in far SE), which also have bluish eyes.

Identification: Medium-sized; eyes green or (in S) bluish; **yellow bar** on 'forehead' (vertex); yellow 'collar' across front of thorax, where curved upper black shoulder stripes usually **meet** black line at rear; contorted line across centre of side of thorax often broken; abdomen has large yellow markings on bases of S3–7, smaller on S8–9; legs mainly black. Slightly smaller and typically darker than Large Pincertail (*p. 196*). ♂ Usually **three** cells in anal triangle (at base of hindwing) (four on Large Pincertail); from above, dark or yellowish tips of upper appendages **overlap**; from side, lower appendage has **knob** close to tip. ♀ Yellow patches on abdomen more extensive than in male; small **yellow protuberance** just behind each eye (absent on Large Pincertail). VARIATION: Yellow 'collar' across front of thorax may be incomplete (as on Large Pincertail). In S, legs mainly yellow; in N, yellow 'forehead' bar may be faint, black shoulder stripes may converge and black more extensive on body, including complete lines across side of thorax.

Behaviour: Sits conspicuously on rocks, short stems or bare ground, with abdomen characteristically arcing to a swollen tip. Often seen on roads or tracks away from water.

Breeding habitat: A range of clear, often rocky rivers and streams, occasionally large, well-oxygenated lakes. Larvae inhabit finer, more silty sediments than Large Pincertail.

LOOK-ALIKES (in range)

Other pincertails (*pp. 196–203*)
Clubtails (*pp. 178–191*)
Green Snaketail (*p. 192*)
Orange-spotted Emerald (*p. 248*)
♀ skimmers (*pp. 262–281*)

♂ ssp. *albotibialis*

♂ anal triangle usually has three cells

♂ ssp. *forcipatus*

♂ Away from Mediterranean, look for green eyes and mainly dark tip to abdomen, including upper appendages.

♂ ssp. *unguiculatus*

♂ + ♀ Southern **subspecies** *albotibialis* and *unguiculatus* have reduced black, with broken line across centre of thorax side.

♂

♂ from above, tips of upper appendages overlap; lower appendage has knob at tip

♂ + ♀ yellow 'forehead' bar

♀

♀

LC Large Pincertail

Onychogomphus uncatus ×1

Blue-eyed Hooktail
Blue-eyed Hook-tailed
Dragonfly

Fairly common =

Stream, river

J F M A M J J A S O N D

| Overall length: | 50–53 mm |
| Hindwing: | 29–33 mm |

♂ + ♀ thorax: complex
continuous black lines across
side between legs and wing
bases

Marginally larger than very similar Small Pincertail (*p. 194*), but less common and widespread and found typically at smaller, faster-flowing waters. Overlaps in S with yellower forms of Small Pincertail and close views or photographs of head, thorax and wing bases are needed to confirm fine details.

Identification: Medium-sized; eyes bluish; 'forehead' (vertex) **all-black**; yellow 'collar' across front of thorax **incomplete**; curved upper black shoulder stripes **do not meet** black line at rear; complex lines on side of thorax link legs to wing bases (central line broken on Small Pincertail in S); abdomen has large yellow markings on bases of S3–7, smaller on S8–9; legs mainly black. Averages slightly larger and yellower than Small Pincertail. ♂ Usually **four** cells in anal triangle (at base of hindwing) (three on Small Pincertail); from above, tips of **bright yellow** upper appendages **do not overlap**; from side, lower appendage **lacks knob** at tip. ♀ Lacks protuberances behind eyes (present in Small Pincertail).

Behaviour: Males sit on rocks, gravel, sand and sometimes vegetation in, over and beside breeding waters. Both sexes are often found away from water.

Breeding habitat: Compared with Small Pincertail, tends to occur in smaller, shadier, faster-flowing, well-oxygenated waters, where larvae inhabit coarse sandy sediments, including upper parts of catchments and fast-flowing parts of larger rivers.

LOOK-ALIKES (in range)

Small Pincertail (*p. 194*)
Clubtails (*pp. 178–191*)
Orange-spotted Emerald (*p. 248*)
♀ skimmers (*pp. 262–281*)

♂ anal triangle usually has four cells

♂ from above, tips of upper appendages do not overlap; lower appendage with no knob at tip

♂ + ♀ 'forehead' black; yellow 'collar' across front of thorax incomplete

EN Faded Pincertail

Onychogomphus costae ×1

Identification of pincertails *p. 177*

Rare and local ▽

River, stream

J F M A M J J A S O N D

Overall length:	43–46 mm
Hindwing:	22–27 mm

♂+♀ thorax: buff and brown,
no black; two inconspicuous whitish stripes on side

A highly localized, inconspicuous clubtail that is restricted to a few flowing waters in Iberia and NW Africa, where it is threatened by poor water quality, abstraction and modifications to watercourses. Its cryptic coloration resembles that of a female skimmer (*pp. 262–281*).

Identification: Small and cryptically coloured in shades of buff and brown with **very little black**; two inconspicuous whitish stripes on side of thorax; abdomen brownish with pale segment bases and black limited to thin lines and small spots; wing-spots **pale orange**; eyes and legs **pale**, the former almost opalescent, the latter with small dark markings. ♂ Abdomen very slender but 'clubbed' at S7–9; appendages long, slender and **pale**, upper almost straight when seen from side, lower upcurved and pointed. ♀ Most obvious markings on abdomen are thick dark streaks along top of S7–9.

Behaviour: Typically found within 100 m of watercourses, including those that have dried up. Perches on or close to the ground, where very well camouflaged amongst low, dry vegetation.

Breeding habitat: Flowing waters with sand or gravel in dry lowlands. Tolerates brackish water and both torrential and intermittent flows; able to live in residual pools in river bed.

LOOK-ALIKES (in range)
Other pincertails (*pp. 194–203*)
Green Hooktail (*p. 204*)
♀ skimmers (*pp. 262–281*)

♂+♀ eyes pale

♂ appendages long, slender and pale – unlike any other dragonfly in range

♂+♀ thorax with two inconspicuous whitish stripes on side

♂+♀ wing-spots pale orange

♂+♀ overall patterning similar to other pincertails but colouring very subdued, with only small black markings

♂+♀ legs pale

NE Waved Pincertail

Onychogomphus flexuosus ×1

Rare and local ?

River

J F M A M J J A S O N D

Overall length:	41–46mm
Hindwing:	25–30mm

♂ + ♀ thorax: pale, with thin
black lines, those on top
enclosing two oval shapes

A small, pale-coloured pincertail that occurs sparsely along gravelly rivers from SW Turkey eastwards.

Identification: Small; thorax with thin black lines, on top enclosing two oval shapes; slender abdomen with alternating broad and narrow black bands on S3–6, producing **two pale yellow bands per segment**; S7–10 mostly **orange**; eyes dull grey-green; wing-spots yellow, with **thick dark borders**; legs yellow at bases. ♂ Abdomen 'clubbed' at S7–10; lower appendage thin and distinctively **wavy** when viewed from the side, upper appendages longer, thin and strongly curved down at tips. ♀ Black lines on top of thorax thinner than on male, otherwise patterning very similar. VARIATION: Extent of black on body may vary; thorax may become greenish.

Behaviour: As other pincertails, may be found perched in low vegetation or on dry ground a short distance from water.

Breeding habitat: Unshaded rivers with gravel banks in hot, dry areas. Records from SW Turkey have come from the Eşen Çayi, near Kadiköy, SE Muğla province, in the same area as Syrian Clubtail (*p. 190*).

LOOK-ALIKES (in range)

Small Pincertail (*p. 194*)
Bladetail (*p. 206*)
♀ skimmers (*pp. 262–281*)

♂

♂

♂ lower appendage wavy

♂+♀ tip of abdomen appears distinctly orange

♀

♀

♂+♀ S3–6 with double bands of black and yellow

♂+♀ legs yellow at base

♂

♀

♂+♀ thick black border to pale wing-spots

NE Dark Pincertail

Onychogomphus assimilis ×1·

Local **?**

River, stream

J F M A M J J A S O N D

Overall length:	50–55mm
Hindwing:	31–34mm

♂ + ♀ thorax: black shoulder stripes usually just touch; broad black stripe across side

A large, dark-looking pincertail that inhabits cool, rocky rivers from SW Turkey eastwards.

Identification: The largest and blackest pincertail, with bluish eyes and a rear end that appears orange; two broad black shoulder stripes just touch part-way along their length; top of thorax **without enclosed yellow ovals** (present in Small Pincertail (*p. 194*)); **broad** black stripe across side of thorax; all-**black** between blue-green eyes (yellow bar in Small Pincertail); broad yellow band across tops of S3–7. ♂ Eyes blue; appendages **orange** and straight when viewed from side, before bending down sharply near tip; upper appendages much longer than lower appendage, which has **bulge halfway** along upper surface. ♀ Eyes blue-green; yellow areas paler than in male. VARIATION: May be a slight gap between black shoulder stripes, especially in female; thorax may be pale greenish in male.

Behaviour: Like other pincertails, males have territorial perches on rocks, often within watercourse, and especially on overhanging waterside branches.

Breeding habitat: Rocky, fast-flowing, cold rivers and streams in mountain foothills, often with wooded margins.

LOOK-ALIKES (in range)
Small Pincertail (*p. 194*)

♂+♀ black shoulder stripes usually just touch (see *opposite*)

♂+♀ broad yellow bands across S3–7

♂+♀ tip of abdomen appears distinctly orange, especially in ♂

♂ appendages large and orange, lower with bulge halfway along

LC Green Hooktail

Paragomphus genei ×1·

Scarce and local =

River, lake, pond

J F M A M J J A S O N D

Overall length:	37–50 mm
Hindwing:	21–26 mm

The smallest of the Gomphidae in Europe, occurring in SW Iberia, Sicily, Sardinia and Corsica, although it is widespread across much of Africa. Males have distinctive combination of a poorly marked green 'front-end' and long, 'hooked' upper appendages.

Identification: Both sexes have **green 'face' and thorax**, the latter marked **weakly** with brownish lines; abdomen has a complex pattern of brown, yellow and black blotches and lines, darkest at tip; eyes blue-grey. ♂ Abdomen slender, reddish-brown and broader towards tip with **'flaps'** on side of S8–9; upper appendages thin, pale and **strongly downcurved**, twice the length of lower appendage, which are dark and upcurved; wing-spots greenish with thick, dark edge. ♀ Thorax and wing-spots duller than on male and abdomen cylindrical, with more extensive yellow markings, only slightly broader at tip. VARIATION: Dark markings variable in extent.

Behaviour: Often sits on ends of twigs, including in treetops when away from water, with abdomen raised towards the sun, in obelisk position.

Breeding habitat: A wide range of slow-flowing and standing waters, often with bare sand or gravel margins and bottom, and rivers with well-vegetated banks. Will frequent pools in otherwise dry riverbeds. Has spread using new artificial sites, such as cattle ponds and reservoirs.

LOOK-ALIKES (in range)

Clubtails (pp. 178–191)
Pincertails (pp. 194–203)
Bladetail (p. 206)
Orange-spotted Emerald (p. 248)

♂ conspicuous 'flaps' on side of S8–9

♂ has distinctive combination of green 'front-end' and long 'hooked' upper appendages

♀ abdomen more-or-less parallel sided

Green Snaketail (*p. 192*) also has green 'front-end' and similar abdominal patterning, but the ranges of the two species do not overlap.

VU Bladetail

Lindenia tetraphylla ×1

PROTECTED

Uncommon ▽

Lake, river

J F M A M J J A S O N D

| **Overall length:** | 69–80 mm |
| **Hindwing:** | 36–40 mm |

This spectacular insect is the largest of the Gomphidae in Europe and the only member of its genus. The 'clubbed' tail is emphasized by conspicuous side 'flaps'. May be locally common in south-east Europe, but true distribution clouded by records of migrants away from confirmed breeding sites. Appears to be spreading, with, for example, breeding confirmed in Bulgaria for the first time in 2017.

Identification: In both sexes, thorax well-patterned in black, with three stripes on each side and two ovals on top; **long, slender, drooping abdomen** has variable amounts of fine black, spiny specks and more extensive black markings, with S8–10 mostly black on top; side of S7–8 adorned with large, brown **'flap'**; wing-spots long and pale or orange, with thick, black borders; eyes olive above, blue-grey below. ♂ Long, **straight** upper appendages. ♀ Base of abdomen distinctly swollen and patterned in orange-brown. VARIATION: Immatures black and pale yellow, duller with age, some becoming pruinose blue and some (in Croatia and Montenegro, at least) largely blackish.

Behaviour: Hunts from perches low down on tracks, rocks, plant stems or wire fences. Abdomen often held raised, with tip drooped, but adopts obelisk position in hot conditions. As with other dragonflies, can be approached closely with care when it is eating prey. Disperses far from water to mature and shows greater migratory tendency than other species of Gomphidae.

Breeding habitat: Lakes and large, slow-flowing rivers, often fringed by reeds. Can also be found at gravel pits and reservoirs with little vegetation. At risk from desiccation and degradation of lakes due to climate change, pollution or abstraction.

LOOK-ALIKES (in range)
Clubtails (*pp. 178–191*)
Pincertails (*pp. 194–203*)
Green Hooktail (*p. 204*)
Slender Skimmer (*p. 280*)

♂ + ♀ Some individuals may be very dark, at least partly due to pruinescence.

♂ + ♀ The combination of separated eyes, conspicuous 'flaps' on the bulging tip of the abdomen and pruinescence is unique in European dragonflies.

♀ base of abdomen swollen, marked in warm brown

GOLDENRINGS, CRUISER & CASCADER

FAMILIES | **Cordulegastridae, Macromiidae & Libellulidae**

GENERA | *Cordulegaster, Macromia & Zygonyx*   9 SPECIES

Large to very large 'fliers' (mean lengths 55–92 mm) with eyes meeting at a point (goldenrings) or broadly. In three disparate genera, they all have a dark body with yellow markings, notably bands (or the appearance of bands) across abdomen. Associated mainly with flowing water, over which males patrol for long periods. Hang more-or-less vertically when perched. Eggs laid in flight (Cascader occasionally lays when settled), sometimes with males in attendance.

Splendid Cruiser and Ringed Cascader are included here because of their similar appearance and habitat use to Goldenrings.

Key features of hawkers, spectres & emperors

◆ large–very large 'fliers'
◆ dark body with yellow markings
◆ yellow 'bands' on abdomen
◆ associated mainly with flowing water
◆ perch hanging vertically

Identify 'sub-types' by eye colour | extent and shape of yellow markings | abdomen shape | presence of long ovipositor | egg-laying habit

Key identification features

SOMBRE GOLDENRING ♂
Cordulegaster bidentata

EYE: colour

EYES: extent of contact

THORAX:
overall colour; patterning on top and side

ANAL TRIANGLE:
3 or 5 cells

ABDOMEN:
overall shape; extent and patterning of yellow markings

BLACK BAR ACROSS FRONS:
strength

OCCIPITAL TRIANGLE:
colour

COMMON GOLDENRING ♂
Cordulegaster boltonii

Goldenrings
7 species

GENUS: *Cordulegaster* (*pp. 212–225*)

KEY FEATURES:

◆ very large
◆ eyes green or blue-green, narrow point of contact (brown in immatures)
◆ yellow bands on abdomen and stripes on thorax
◆ abdomen slightly swollen near tip
◆ long ovipositor extends beyond abdomen tip (female)
◆ breed in flowing waters
◆ eggs laid in flight, vertically with repeated stabbing action

Identify species by eye colour | shape of male appendages | strength of black bar across frons | colour of occipital triangle | number of cells in male anal triangle | pattern of yellow on thorax and abdomen (extent of yellow on mid-panel of side of thorax; shape of yellow marking on side of S1; extent and shape of yellow on top and side of abdomen) | location

Goldenring

Cruiser
1 species

GENUS: *Macromia* (*p. 226*)

KEY FEATURES:

◆ large
◆ eyes green, broad contact
◆ metallic green thorax
◆ abdomen slightly swollen near tip
◆ very long legs
◆ breeds in slow-flowing or standing waters
◆ eggs laid in flight, dipping into surface

Cascader
1 species

GENUS: *Zygonyx* (*p. 228*)

KEY FEATURES:

◆ large (but smaller than rest of group)
◆ eyes brown, broad contact
◆ yellow markings only on side of thorax and abdomen
◆ abdomen with parallel sides, not swollen near tip
◆ breeds in waterfalls and rapids
◆ lays eggs in flight in tandem, or by female alone in flight or when settled

Cruiser

Cascader

Goldenrings compared

Pointers indicate important identification features included in the table (*opposite*) that are visible from above.

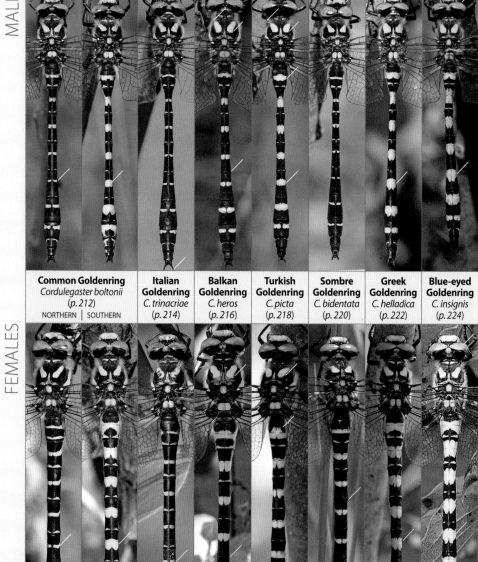

| **Common Goldenring** *Cordulegaster boltonii* (*p. 212*) NORTHERN \| SOUTHERN | **Italian Goldenring** *C. trinacriae* (*p. 214*) | **Balkan Goldenring** *C. heros* (*p. 216*) | **Turkish Goldenring** *C. picta* (*p. 218*) | **Sombre Goldenring** *C. bidentata* (*p. 220*) | **Greek Goldenring** *C. helladica* (*p. 222*) | **Blue-eyed Goldenring** *C. insignis* (*p. 224*) |

Goldenrings compared (*Cordulegaster* species)

Feature	Common Goldenring *C. boltonii* (p. 212)	Italian Goldenring *C. trinacriae* (p. 214)	Balkan Goldenring *C. heros* (p. 216)	Turkish Goldenring *C. picta* (p. 218)	Sombre Goldenring *C. bidentata* (p. 220)	Greek Goldenring *C. helladica* (p. 222)	Blue-eyed Goldenring *C. insignis* (p. 224)
Eye colour	Green				Green		Blue-green
Black bar across frons	Small	Small (females) or trace	Faint (males) or clear (females)	Variable	Wide	None or small	
Occipital triangle	Yellow		Black, may have 2 small yellow spots	Black with 2 small yellow spots	Black	Usually black	Yellow
Extent of yellow on mid-panel on side of thorax	Large 'exclamation mark' ('!')		2 or 3 irregular marks	Large 'exclamation mark' ('!'), some with 3 marks	Small or absent	Usually small, tapering from top	Small, tapering from top
Yellow marking on side of S1	arc on rear/lower edge				oblique mark near top		
Extent of wide yellow bands on top of abdomen	15% (NORTHERN) 25% (SOUTHERN)	Less than 20%	About 25%		About 20%	About 40%	50–60%
Segment tips with paired spots	S2–8	S2–6	S2–4	S2–4/5/6	S2–3/4/5	S2–4/5/6	
Shape of yellow on side of abdomen	Parallel or tapering	Triangular	Parallel		Triangular	Tapering	Parallel
♂ upper appendages	1 'tooth' on underside				2 'teeth' on underside		
♂ anal triangle	Usually 5 cells				Usually 3 cells		
Range	Widespread, not SE	S. Italy	SE	Far SE	Central and SE	S Greece	Far SE

[LC] **Common Goldenring**

Cordulegaster boltonii ×1

Golden-ringed Dragonfly, Golden-ringed Spiketail

Fairly common =

River, stream, flush

J F M A M J J A S O N D

Overall	MALE 74–80 mm
length:	FEMALE 80–85 mm
Hindwing:	MALE 40–47 mm
	FEMALE 45–51 mm

♂ upper appendages: =S10, diverge, almost touch at base

A distinctive, large and impressive black-and-yellow dragonfly with green eyes, the female having a long, spike-like ovipositor (hence the alternative name 'spiketail' for this genus). It is the most widespread European goldenring and the only species in much of the N and W, being replaced by other species in SE.

Identification: Sexes similar, with green eyes; black bar across frons small; occipital triangle yellow; thorax side with relatively large yellow stripe between two broader yellow stripes, often forming an 'exclamation mark' ('!'); side of S1 with yellow **arc on lower/rear edge**; yellow 'rings' across abdomen cover 15% of total length; paired yellow lines on tips of S2–**8**. ♂ Anal triangle usually has **five** cells; upper appendages as long as S10, from above diverging but almost touching at base, each with **one** 'tooth' on underside; lower appendage from below broad, parallel-sided and tip only slightly notched. VARIATION: Eyes brown in immatures; variants in Spain, S France and NW Italy (formerly subspecies *immaculifrons* and *algirica*, now invalidated by DNA studies) have significantly more yellow on the abdomen (covering 25% of the total length), especially on S3–4; in Italy, the occipital triangle may be black with, on some, a pair of tiny yellow dots.

Behaviour: As with other goldenrings, males patrol long territories at suitable breeding waters, flying low and rather slowly, hovering frequently and passing close to patient observers. Males often perch on waterside vegetation between territorial patrols and both sexes may wander far from water.

Breeding habitat: Flowing waters, often in uplands, woodland and heathland. Sites range from springs and boggy runnels to small rivers, where sand, silt or peat debris has been deposited in quieter stretches.

LOOK-ALIKES (in range)

Other goldenrings (*pp. 214–225*)
Splendid Cruiser (*p. 226*)
Ringed Cascader (*p. 228*)

♂

♂ + ♀ occipital triangle yellow

♂ + ♀ yellow 'exclamation mark' ('!') on side of thorax

♂

♂ anal triangle usually has five cells (rarely 3–8)

♂ + ♀ yellow on abdomen more extensive in S Europe

♂ Southern European variant

♀

♀

♂ + ♀ yellow mark on rear edge of S1

♀ larger than ♂, usually with more extensive yellow markings

E NT **Italian Goldenring**

Cordulegaster trinacriae ×1

Goldenrings compared *pp. 210–211*

PROTECTED

Scarce and local ▽

River, stream

J F M A M J J A S O N D

Overall length:	MALE 73–79 mm
	FEMALE 83–93 mm
Hindwing:	MALE 45–47 mm
	FEMALE 51–53 mm

Endemic to S Italy and Sicily, overlapping in the north of its range with the almost identical Common Goldenring (*p. 212*), from which it was first differentiated as recently as 1976. Separation is best confirmed through in-hand examination of male appendages.

Identification: Sexes similar, with green eyes; occipital triangle yellow; thorax side with yellow stripe between two broader yellow stripes that, on some, forms an 'exclamation mark' ('!'); side of S1 with yellow **arc on lower/rear edge**; yellow 'rings' across top of abdomen cover less than 20% of total length (25% in southern variants of Common Goldenring), **tapering** to a point on side; paired yellow lines on tips of S2–**6**. ♂ Frons with only trace of black bar; anal triangle usually has **five** cells; upper appendages slightly longer than S10, from above diverging but almost touching at base, each with **one** 'tooth' on underside; from below, rear edge of lower appendage **deeply notched** (only slightly notched in Common Goldenring). ♀ Small black bar across frons. VARIATION: Eyes brown in immatures.

Behaviour: Males are typically found patrolling low over suitable watercourses, but both sexes may wander far from water.

Breeding habitat: Streams and small rivers with sandy bottom, especially in wooded upland areas.

♂ upper appendages: >S10, diverge, almost touch at base

LOOK-ALIKES (in range)
Common Goldenring (*p. 212*)
Sombre Goldenring (*p. 220*)

♀

♀

♂

♂

♂ + ♀ yellow markings on abdomen taper to a fine point on the side

♂ + ♀ yellow markings on abdomen slightly smaller than on Common Goldenring (*p. 212*) in S part of that species' range

♂ anal triangle usually has five cells (rarely 3–8)

♂ lacks dark bar across frons; occipital triangle yellow

ⓔ NT Balkan Goldenring

Cordulegaster heros ×1

PROTECTED

Scarce and local =

River, stream

J F M A M J J A S O N D

Overall length:	MALE 77–84 mm
	FEMALE 88–96 mm
Hindwing:	MALE 45–50 mm
	FEMALE 53–58 mm

♂ upper appendages: =S10, diverge, almost touch at base

Europe's largest goldenring; endemic to the SE, where it replaces the very similar Common Goldenring (*p. 212*). However, it is sometimes found with Sombre Goldenring (*p. 220*), and rarely with Common Goldenring in the N of its range in Austria. It was recognized as a distinct species as recently as 1979 and the N and E limits of its range are still uncertain. Identification is best confirmed through careful examination of critical features.

Identification: Sexes similar, with green eyes; occipital triangle black; yellow shoulder stripes have an **angular** outer corner at the rear, some with a small yellow spot nearby; thorax side with two or three irregular yellow marks between two broader stripes; side of S1 with yellow **arc on lower/rear edge**; yellow 'rings' on abdomen cover about 25% of total length (*i.e.* more than on Common and Sombre Goldenrings that occur in the range of this species); paired yellow lines on tips of S2–4. ♂ Frons with faint black bar; anal triangle usually has **five** cells; upper appendages as long as S10, from above diverging but almost touching at base, each with **one** 'tooth' on underside. ♀ Clear black bar across frons.

VARIATION: Eyes brown in immatures; on some, especially females, the occipital triangle has two small yellow spots; clear black bar across frons in males of ssp. *pelionensis* in S of range (Greece, Albania, Bulgaria).

Behaviour: Males fly low along suitable watercourses, while both sexes may be found hanging in the shade on the edge of nearby woodland.

Breeding habitat: Shady streams and small rivers with sandy bottom, especially in wooded upland areas, tending to avoid headwaters where Sombre Goldenring occurs.

LOOK-ALIKES (in range)

Common Goldenring (*p. 212*)
Turkish Goldenring (*p. 218*)
Sombre Goldenring (*p. 220*)
Greek Goldenring (*p. 222*)
Blue-eyed Goldenring (*p. 224*)

♀

♀i

♂

♂

♂+♀ yellow
shoulder stripes
have sharp angle
on outer corner
of rear edge

♂+♀ side of
thorax with 2–3
irregular marks
between yellow
stripes

♂ anal triangle
usually has five
cells (rarely 3–8)

♂+♀ occipital triangle black;
♂ ssp. *pelionensis* [*above*] shows
clear black bar across frons

VU **Turkish Goldenring**

Cordulegaster picta ×1

Scarce and local	=	
River, stream		

J F M A M J J A S O N D

Overall length:	MALE	72–80 mm
	FEMALE	80–89 mm
Hindwing:	MALE	43–46 mm
	FEMALE	48–53 mm

♂ upper appendages: =S10, diverge widely, almost touch at base

Found from the S Balkans eastwards, where its range overlaps that of Blue-eyed Goldenring (*p. 224*) and potentially Sombre Goldenring (*p. 220*) and Balkan Goldenring (*p. 216*). The extent of the yellow patterning is very variable in this species and identification is best confirmed through close examination of male appendages.

Identification: Sexes similar, with green eyes; occipital triangle black with two yellow patches; yellow shoulder stripes have **rounded** outer corner at rear, some with yellow spots nearby; thorax side with large yellow 'exclamation mark' ('!') between two broader yellow stripes that may be broken into three sections; side of S1 with yellow **arc on lower/rear edge**; yellow 'rings' on abdomen cover about 25% of total length (typically more than in Sombre Goldenring); paired yellow lines on tips of S2–4/5/6. ♂ Anal triangle usually has **five** cells; upper appendages as long as S10, from above **diverging widely from base**, each with **one** 'tooth' on underside. ♀ Extensive yellow marking on top of S2 often extends forwards (not so in Balkan Goldenring). VARIATION: Eyes brown in immatures; black bar across frons very variable, absent on some.

Behaviour: Like other goldenrings, males patrol low along the margins of long territories, searching for females. Both sexes feed around nearby woodland and rest in trees and shrubs.

Breeding habitat: Streams and medium-sized rivers with sandy bottom, often shaded and in upland areas. May occur with Blue-eyed Goldenring at springs and seepages.

LOOK-ALIKES (in range)

Balkan Goldenring (*p. 216*)
Sombre Goldenring (*p. 220*)
Blue-eyed Goldenring (*p. 224*)

♂ + ♀ yellow shoulder stripes have rather rounded angle on outer corner of rear edge

♂ + ♀ side of thorax with yellow 'exclamation mark' ('!') between yellow stripes

♀

♀

♂

♂

♂ anal triangle usually has five cells (rarely 3–8)

♂ + ♀ occipital triangle black, with two yellow spots

Ⓔ ☒ Sombre Goldenring

Cordulegaster bidentata ×1

Two-toothed Golden-ringed Dragonfly

| Local ▽ |
| Stream, flush |

J F M A M J J A S O N D

Overall length:	MALE 69–78mm
	FEMALE 74–83mm
Hindwing:	MALE 41–46mm
	FEMALE 45–50mm

♂ upper appendages: >S10, parallel, wide apart at base

A small and rather dark endemic goldenring that is fairly widespread, although most abundant in the SE, where populations generally have more extensive yellow patterning. There is apparently no overlap with Greek Goldenring (*p. 222*) or Blue-eyed Goldenring (*p. 224*) that occur to the SE, and it is not usually found with Common Goldenring (*p. 212*).

Identification: Sexes similar, with green eyes; frons with **wide** black bar; occipital triangle black; thorax side with **little or no** yellow between two broader yellow stripes; side of S1 with **oblique yellow patch near top**; yellow 'rings' across abdomen cover about 20% of total length; paired yellow lines on tips of S2–3/4/5, 'rings' extending as 'wedges' onto sides of S2. ♂ Anal triangle usually has **three** cells; upper appendages slightly longer than S10, from above **parallel and wide apart at base**, each with **two** 'teeth' on inner half of underside; lower appendage from below **tapering** to rear. VARIATION: Eyes brown in immatures; yellow on centre of thorax side may form small 'wedge' or 2–3 spots; yellow on abdomen can be more extensive in S/SE; occipital triangle may be partly yellow in far SE.

Behaviour: As with other goldenrings, males patrol long territories at suitable breeding waters, flying low and rather slowly. Both sexes may wander far from water to feed and rest.

Breeding habitat: Small streams, runnels, springs, steep flushes and rocky seepages, often calcareous (alkaline) and typically in headwaters in open upland woodland. Watercourses therefore tend to be smaller than those used by Common Goldenring.

LOOK-ALIKES (in range)
Other goldenrings (*pp. 212–225*)
Splendid Cruiser (*p. 226*)

♀

♀

♂

♂

♂ + ♀ wide black
bar across frons

♂ + ♀ abdomen
with restricted
yellow markings

♂ + ♀ side of
thorax with little
or no yellow
between broad
yellow stripes

♂ anal triangle
usually with
only three cells

E EN Greek Goldenring

Cordulegaster helladica ×1

Rare and local ▽

Stream, spring

J F M A M J J A S O N D

Overall length:	MALE 68–78 mm
	FEMALE 78–83 mm
Hindwing:	MALE 41–46 mm
	FEMALE 46–49 mm

♂ upper appendages: =S10, parallel, wide apart at base

This small goldenring is endemic to southern Greece. The yellow patterning on the abdomen is very similar to that of Blue-eyed Goldenring (*p. 224*), which occurs to the E, but more extensive than in Balkan Goldenring (*p. 216*), which occurs in the same region.

Identification: Sexes similar, with green eyes; frons with black bar **small or absent**; sharply angled upper outer corners of shoulder stripes; thorax side with small 'wedge' of yellow between two broader yellow stripes; side of S1 with **oblique yellow patch near top**; yellow 'rings' across abdomen cover about **40%** of total length; paired yellow lines on tips of S2–4/5/6, 'rings' extending as broad 'wedges' down sides of S2–7. ♂ Anal triangle usually has **three** cells; upper appendages from above as long as S10, **parallel and wide apart at base**, each with **two** 'teeth' on inner half of underside; lower appendage from below **tapering** to rear. VARIATION: Eyes brown in immatures; occipital triangle black or yellow; yellow on centre of thorax side forms large rectangular panel at some locations; reduced amounts of yellow on abdomen on Cyclades Islands.

Behaviour: As with other goldenrings, males patrol low along territories at suitable breeding waters. Both sexes may be found around nearby scrub and woodland.

Breeding habitat: Upper reaches of small rocky streams in hilly forests and scrub within woodland. At Delphi, occurs in the karstic outflow at the Castalian spring, between the Phaedriades. Elsewhere, known mainly from the Peloponnese, Euboea Island, Attica and the Cyclades Islands (Andros, Tinos and Naxos).

LOOK-ALIKES (in range)
Balkan Goldenring (*p. 216*)

♂+♀ no or only small black bar across frons

♂+♀ yellow on centre of thorax side typically forms a 'wedge' shape

♀

♀

♂+♀ abdomen with extensive yellow markings

♂

♂

♂ anal triangle usually with only three cells

EN Blue-eyed Goldenring

Cordulegaster insignis ×1

Uncommon ?
Stream, spring

J F M A M J J A S O N D

Overall length:	MALE 71–78mm
	FEMALE 71–83mm
Hindwing:	MALE 40–46mm
	FEMALE 41–49mm

♂ upper appendages: =S10, parallel, wide apart at base

A colourful goldenring, found in the far SE of Europe. The yellow patterning on the abdomen is generally more extensive than on other goldenrings that occur in this part of the region and this, in combination with the blue eyes, draws comparison with Magnificent Emperor (*p. 174*) as much as a goldenring.

Identification: Sexes similar, with **blue-green** eyes; frons with black bar **small or absent**; occipital triangle **yellow, swollen at rear**; thorax side with small 'wedge' of yellow between two broader yellow stripes; side of S1 with **oblique yellow patch from top front**; yellow 'rings' across abdomen cover **50–60%** of total length, extending as broad 'wedges' down sides of S2–6/7; paired yellow lines on tips of S2–4/5/6. ♂ Anal triangle usually has **three** cells; upper appendages from above as long as S10, **parallel and wide apart at base**, each with **two** 'teeth' on inner half of underside; lower appendage from below **parallel or slightly tapering** to rear. VARIATION: Eyes brown in immatures; a large, green-eyed form occurs on Greek island of Ikaria.

Behaviour: Males fly patrol regularly along territories at suitable breeding waters, flying low in search of females.

Breeding habitat: Upper reaches of shady streams; also seepages and artificial channels. Generally smaller streams than those used by Turkish Goldenring (*p. 218*), often rocky with sandy beds.

LOOK-ALIKES (in range)
Turkish Goldenring (*p. 218*)
Magnificent Emperor (*p. 174*)

♀

♀

♂+♀ no or only small black bar across frons

♂+♀ occipital triangle yellow

♂+♀ abdomen with extensive yellow markings

♂+♀ eyes blue-green

♂

♂

♂+♀ side of thorax typically with yellow 'wedge' shape at centre

♂ anal triangle usually with only three cells

ⓔ 🆅🆄 **Splendid Cruiser** *Macromia splendens* ×1

PROTECTED

Rare and local ▽

River, lake

J F M A M J J A S O N D

| Overall length: | 70–75 mm |
| Hindwing: | 42–49 mm |

This spectacular river cruiser, the only European representative of the widespread family Macromiidae, is placed here because its green eyes and bold black-and-yellow patterning give more than a passing resemblance to a goldenring (*pp. 212–225*). It is endemic to SW Europe, occurring mainly in SW France and W Iberia, and is very elusive and hard to find.

Identification: A large 'flier', both sexes of which have green eyes that have a broad zone of contact, **metallic dark green** thorax with yellow shoulder stripes, **crescent** in front of wing bases and a **single** stripe across side of thorax; legs very long. ♂ Abdomen slender, bulging towards tip, black with yellow bands across tops of S2–4/5 and S7–8, largest patch on S7. ♀ Abdomen parallel-sided, black with variable yellow markings on top of S2–7. **VARIATION:** Yellow spots on abdomen variable in size.

Behaviour: Males patrol for long periods over watercourses fast and at about knee height (higher than goldenrings), with abdomen raised and drooping at the tip. They keep a metre or more from the bank and aggressively pursue other dragonflies, including Blue Emperor. Rarely seen perched. Eggs are laid alone in flight by dipping abdomen tip into water (unlike the vertical egg-laying by goldenrings).

Breeding habitat: Calm, including dammed, stretches of unpolluted, warm rivers, often with rocks and overhanging trees and shrubs along the edges; can also be found at deep pools in smaller, temporary watercourses and hydroelectric reservoirs with fluctuating water levels and bare margins.

LOOK-ALIKES (in range)
Common Goldenring (*p. 212*)
Sombre Goldenring (*p. 220*)
Ringed Cascader (*p. 228*)

♂+♀ single yellow stripe on side of thorax

♀

♀

♂+♀ large yellow markings on S3 and S7 (especially noticeable in flight)

↗

♂

♂+♀ top of thorax metallic green with yellow stripes

♂+♀ yellow crescent in front of wing bases

Ⓥ Ringed Cascader

Zygonyx torridus ×1

Scarce and local ▽

River, stream

J F M A M J J A S O N D

| Overall length: | 50–60 mm |
| Hindwing: | 45–50 mm |

Although shorter than the preceding goldenrings (*pp. 212–225*) and Splendid Cruiser (*p. 226*), this unrelated species is dealt with here because of its basic black-and-yellow patterning and its association with fast-flowing water. It is the largest member of the Libellulidae (a large family that includes chasers, skimmers and darters) and its long wings allow it to patrol for long periods, like a hawker, often over waterfalls and other stretches of turbulent water.

Identification: Both sexes shiny and blackish, with **large yellow spots along side** and small yellow marks along top of abdomen, appearing as rings; eyes brown above, grey below, with broad zone of contact. ♂ Thorax with indistinct brown markings, little or no yellow. ♀ Thorax with irregular yellow markings on top and side; yellow markings on abdomen more extensive than on male. VARIATION: Thorax may have a greyish bloom; extent of yellow on abdomen variable, including up to four narrow yellow rings over top of basal segments; wings may have an amber tinge to varying extent.

Behaviour: Males patrol often small sections of fast-flowing and turbulent waters, hovering over open stretches. They perch more-or-less vertically, unlike most species in the family Libellulidae. Eggs are either laid at rest alone, or in flight alone or in tandem – an unusually diverse range of egg-laying behaviours.

Breeding habitat: Warm, permanent rivers and streams, including waterfalls and rapids, especially where the water is shallow. Widespread in Africa, but in Europe confined to the Canary Islands (Gran Canaria, La Gomera, La Palma and Tenerife) and parts of S Iberia; also breeds at one site in Sicily. Prone to wandering far from breeding sites.

Yellow spots on side of abdomen may appear to form rings, as on goldenrings.

LOOK-ALIKES (in range)
Common Goldenring (*p. 212*)
Splendid Cruiser (*p. 226*)

♂+♀ amber tinge to wings on some individuals

♀ more extensive yellow markings than ♂

♂ thorax mainly dark, usually shiny

EMERALDS

FAMILIES | **Corduliidae & uncertain [*incertae sedis*]**

GENERA | *Cordulia, Corduliochlora, Somatochlora & Oxygastra*   9 SPECIES

Medium-sized (mean lengths 48–53 mm), rather elusive 'fliers' with shiny green eyes and largely dark, often metallic green or bronzy thorax and abdomen. The abdomen is slender and conspicuously waisted on males, which have angled inner corners of hindwing. Females have a rather parallel abdomen and rounded inner corners to hindwing. Males patrol territories in persistent, low flight, perching away from water, often in trees. Eggs are laid in flight alone into water or debris at or near the edges of flowing waters, ponds, lakes or bog pools, depending on the species. Both sexes forage around edges and tops of trees, if these are present.

Key features of emeralds

◆ medium-sized 'fliers'
◆ shiny green eyes (when mature)
◆ dark body, often green/bronzy, with metallic sheen
◆ male abdomen waisted, clubbed at tip in some

Identify species by 'face' pattern | extent of metallic green on body | shape of male abdomen | extent of yellow on side of thorax/abdomen | presence of pale 'crest' on S10 | male appendages | female vulvar scale

Emeralds in flight

Emeralds are most often seen in flight; all the species are shown here for comparison.

tip held high

Downy Emerald
Cordulia aenea (p. 232)

'cheeks' yellow

Bulgarian Emerald
Corduliochlora borisi (p. 234)

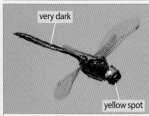

shining emerald green
'U'-shaped yellow

Brilliant Emerald
Somatochlora metallica (p. 236)

shining emerald green
small yellow mark

Balkan Emerald
Somatochlora meridionalis (p. 238)

yellow spots

Yellow-spotted Emerald
S. flavomaculata (p. 240)

very dark
yellow spot

Northern Emerald
Somatochlora arctica (p. 242)

blackish
whitish ring

Alpine Emerald
Somatochlora alpestris (p. 244)

blackish
whitish ring
yellow spot

Treeline Emerald
Somatochlora sahlbergi (p. 246)

pale spot
orange-yellow marks

Orange-spotted Emerald
Oxygastra curtisii (p. 248)

Emeralds compared

Important identification features are indicated.

Species	'Face'	Male appendages		Female vulvar scale	Other key features
		above	side		
Downy Emerald *Cordulia aenea* (p. 232)	dark	LOWER: forked	= length	short and flat	no yellow spots
Bulgarian Emerald *Corduliochlora borisi* (p. 234)	yellow 'cheeks'	LOWER: notched	downcurved	short	S2–3: yellow spots
Brilliant Emerald *Somatochlora metallica* (p. 236)	yellow sides meet			long, pointed	no yellow mark
Balkan Emerald *Somatochlora meridionalis* (p. 238)	yellow sides meet			long, pointed	yellow mark
Yellow-spotted Emerald *Somatochlora flavomaculata* (p. 240)		UPPER: ± straight	curled tip		THORAX AND ABDOMEN: yellow markings on side
Northern Emerald *Somatochlora arctica* (p. 242)		UPPER: 'callipers'	± straight	long, flat / extends ≥ S9	♀ S3: 2 yellow spots
Alpine Emerald *Somatochlora alpestris* (p. 244)		UPPER: double angle	kink near tip	long, triangular / extends ½ S9	2 cross-veins
Treeline Emerald *Somatochlora sahlbergi* (p. 246)		UPPER: bent inwards	kink near tip / hairy	short	1 cross-vein
Orange-spotted Emerald *Oxygastra curtisii* (p. 248)	all-dark		spine near base	short	ABDOMEN: orange-yellow markings along top. S10: pale yellow 'crest'

231

LC Downy Emerald

Cordulia aenea ×1

Emeralds compared *pp. 230–231*

Common =
Lake, pond, canal, river

J F M A M J J A S O N D

Overall length:	47–55 mm
Hindwing:	29–35 mm

♂ + ♀ 'face' dark, yellow
only on mouthparts

One of the more widespread emeralds, this is the only member of its genus in Europe. It is found at sheltered wooded ponds and lakes and flies earlier in the season than other emeralds.

Identification: Medium-sized. Body appears dark except at close range, when **bright green** eyes and downy **bronzy-green** thorax shine in sun (note that all emeralds have more-or-less downy thorax and abdomen); abdomen **darker** bronzy-green, with incomplete narrow yellow ring at base of S3; 'face' dark, lacking yellow between eyes (only mouthparts yellowish); golden bases to otherwise clear wings. ♂ Abdomen markedly **waisted** at S3, **bulging around S7–8** (nearer tip than on most other emeralds) and held above horizontal in flight; rear margin of the hindwing is acutely angled; upper and lower appendages of **equal length**, the lower forked and ending in two uptilted hooks. ♀ Abdomen has almost parallel sides beyond swollen S2; large whitish patch on lower side of S3; vulvar scale short and held flat against S9.
VARIATION: Immatures have brown eyes; yellow ring on S2/3 variable, absent on some.

Behaviour: Males patrol close to the bank, frequently hovering for a few seconds 0·5–1 m above water in small, sunny bays. Rarely lands beside water, usually in trees some distance away. Feeds along woodland edges, typically at canopy height. Eggs laid alone by flicking tip of abdomen into shallow water, often beneath overhanging branches.

Breeding habitat: Usually found at well-vegetated neutral and acidic ponds and lakes, sometimes canals and slow-flowing rivers, often fringed with trees or with woodland nearby. Mostly at mountain lakes in S.

LOOK-ALIKES (in range)
Other emeralds (*pp. 234–249*)

♀ abdomen almost parallel-sided; whitish patch on lower side of S3

♀i

♀

♂ + ♀ thorax downy; bronzy-green

♂ + ♀ eyes bright green

♂

♂ abdomen dark, waisted at S3, club-tipped and raised at tip in flight

VU Bulgarian Emerald

Corduliochlora borisi ×1

Rare and local ▽

River, stream

J F M A M J J A S O N D

Overall length:	45–50 mm
Hindwing:	31–34 mm

♂+♀ yellow 'U' between eyes extends down onto 'cheeks', sometimes visible given good flight views

Discovered in Bulgaria in 1999 by Milen Marinov and named after his son, Boris. Initially placed in the genus *Somatochlora*, this rare species is more similar to Downy Emerald (*p. 232*) than the other emeralds. It is known currently from about 30 locations in SE Bulgaria, NE Greece and European Turkey.

Identification: Medium-sized, but a relatively small emerald with **bright green** eyes; 'face' has **yellow 'U' between eyes** (on frons, as on Balkan Emerald (*p. 238*)) and **extending down onto 'cheeks'** (which are black on other emeralds); thorax **metallic emerald green**; abdomen darker bronzy-green, with pairs of yellow marks on top and sides of S2–3 (larger on female) and incomplete narrow yellow ring around base of S3; wings with yellow suffusion at bases. ♂ Abdomen waisted at S3 and obviously **swollen around S7–8** (as Downy Emerald, S5–7 on other emeralds); tip held less elevated in flight than Downy Emerald; upper appendages blunt, curving down and out at tips (longer than Downy Emerald) with two 'teeth' on underside, lower shorter with upturned tip from side (lacks second hook of Downy Emerald). ♀ Abdomen has parallel sides and additional small yellow markings on side of S4–5, vulvar scale much shorter than S9 and not prominent. VARIATION: Immatures have brown eyes; immatures and females have yellow suffusion across front of wings.

Behaviour: Flight period earlier in the year than other emeralds except Downy Emerald, with few by end of June, when Balkan Emerald becomes the most frequent emerald in SE Europe. Territorial males patrol low over water in the shade of trees. Usually feeds around nearby tree canopy.

Breeding habitat: Shaded, small rivers and streams with areas of calm water in woodland.

LOOK-ALIKES (in range)

Downy Emerald (*p. 232*)
Balkan Emerald (*p. 238*)
Yellow-spotted Emerald (*p. 240*)

♀ i

♂ + ♀ S2 & S3 with pair of large yellow spots

♀ i

♂

♂ abdomen obviously waisted at S3 and clubbed at S7–8, as Downy Emerald (*p. 232*)

LC Brilliant Emerald

Somatochlora metallica × 1·

Common =
Lake, pond, canal, river

J F M A M J J A S O N D

Overall length:	50–55 mm
Hindwing:	34–38 mm

♂ + ♀ yellow 'U' between eyes

This striking emerald is widespread across much of C and NE Europe, being replaced in the SE by the very similar Balkan Emerald (*p. 238*).

Identification: Medium-sized, with **bright green** eyes and **whole body** shining metallic emerald green; 'face' has **yellow 'U' between eyes** (on frons); abdomen longer than on Downy Emerald (*p. 232*), with pairs of small yellow marks on S2–3 (larger on female) and incomplete narrow yellow ring around base of S3; wings with variable yellow **suffusion**, deeper at bases. ♂ Abdomen waisted at S3 and slightly swollen around S5–7 (S7–8 on Downy Emerald); tip held less elevated in flight than Downy Emerald; upper appendages long and slender, converging from base and turning up at the tip (rather short and blunt on Downy Emerald). ♀ Abdomen has parallel sides and **conspicuous 'spike'** (vulvar scale) that is longer than S9 and projects at a right angle below that segment; costa yellow. VARIATION: Immatures have brown eyes and more extensive yellow suffusion across front of wings. Some have a bronzy abdomen; can look dark in some lights.

Behaviour: Territorial males patrol the water's edge, often under trees, keeping slightly higher and farther from shore than Downy Emerald and not pausing to hover as frequently. Usually go to nearby trees or bushes to feed, rest or mate. Females lay eggs alone, repeatedly beating vulvar scale into shady shallow water or damp margins with tip of abdomen angled up.

Breeding habitat: Ponds and lakes at least partially shaded by trees and with woodland close by, relatively deep water and deposits of mud and organic debris. Sometimes small, slow-flowing rivers, canals and peaty moorland lakes; montane lakes in S.

LOOK-ALIKES (in range)
Other emeralds (*pp. 232–249*)

♀ vulvar scale conspicuous 'spike'

♀

♀

♂

♂

♂ + ♀ eyes and body shiny green, can appear dark at some angles

♂ + ♀ wings with variable yellow suffusion, strongest in IMMATURES

♂ abdomen waisted at S3, swollen at S5–7 (S7–8 on Downy Emerald (p. 232)), and held straight

LC Balkan Emerald

Somatochlora meridionalis × 1·5

J F M A M J J A S O N D

Overall length:	50–55 mm
Hindwing:	34–38 mm

♂ + ♀ yellow 'U' between eyes

Replaces the almost identical Brilliant Emerald (*p. 236*) in streams and rivers in SE Europe. Individuals with intermediate features are known from the N of the range.

Identification: Medium-sized, with **bright green** eyes and **whole body** shining metallic emerald green; 'face' has **yellow 'U' between eyes** (on frons); side of thorax with small yellow mark across centre (absent on Brilliant Emerald, but beware of yellowish reflections); abdomen longer than on Downy Emerald (*p. 232*), with pairs of yellow marks on S2–3 (larger on female; larger than on Brilliant Emerald) and incomplete narrow yellow ring around base of S3; wings with variable yellow **suffusion**, deeper at bases, wing-spots dark (Brilliant Emerald brown, but variable). ♂ Abdomen waisted at S3 and slightly swollen around S5–7 (S7–8 on Downy Emerald); tip held less elevated in flight than Downy Emerald; upper appendages long and slender, converging from base and turning up at the tip (slightly shorter on Brilliant Emerald). ♀ Abdomen has parallel sides and **conspicuous 'spike'** (vulvar scale) that is longer than S9 and projects at a right angle below that segment; costa yellow. VARIATION: Immatures have brown eyes and more extensive yellow suffusion across front of wings.

Behaviour: Territorial males patrol long stretches low over shaded water, avoiding sunny areas, hovering frequently; may fly even when cloudy in warm conditions. Both sexes feed around sunny glades and along woodland edges.

Breeding habitat: Breeds almost invariably in shaded streams and rivers in lowlands (Brilliant Emerald often at standing waters, and in upland areas in SE).

LOOK-ALIKES (in range)
Downy Emerald (*p. 232*)
Brilliant Emerald (*p. 236*)
Bulgarian Emerald (*p. 234*)
Yellow-spotted Emerald (*p. 240*)

♀

♀ vulvar scale
conspicuous
'spike'

♂

♂+♀ diagnostic
small yellow mark
on side of thorax

♂

♂ thorax and
abdomen metallic
green, as Brilliant
Emerald (p. 236)

LC Yellow-spotted Emerald

Somatochlora flavomaculata × 1·5

Locally common =

Lake, pond, canal, river

J F M A M J J A S O N D

Overall length:	45–54 mm
Hindwing:	34–43 mm

♂ + ♀ yellow spot on each side of 'face'

Perhaps the most distinctive emerald, with more-or-less obvious yellow markings on the side. It is fairly widespread in lowlands across C Europe, becoming more localized in the S and W.

Identification: Medium-sized, slightly smaller than Brilliant Emerald (*p. 236*); yellow spot on each side of 'face', next to **bright green** eyes, extending towards centre; thorax **metallic emerald green** with two yellow elongated marks down side (but beware reflections on other emeralds); abdomen **blackish with yellow markings** on sides of segments, largest on S2–3, and narrow yellow ring around base of S3 (incomplete in male); wings with yellow **suffusion** at bases. ♂ Abdomen waisted at S3 and slightly swollen around S5–7 (S7–8 in Downy Emerald (*p. 232*)); tip held less elevated in flight than in Downy Emerald; upper appendages from above smooth-sided and straight for most of length, turning up at the tip (rather short and blunt in Downy Emerald). ♀ Abdomen has parallel sides and obvious 'spike' (vulvar scale) **shorter** than S9 (longer in Brilliant Emerald and Balkan Emerald (*p. 238*)), held out conspicuously below S9; costa yellow. **VARIATION:** Immatures have brown eyes and bright yellow body markings. Yellow markings smaller on males, becoming darker with age in both sexes and especially hard to see in old males.

Behaviour: Territorial males patrol, often at eye-level, over breeding habitats, but adults frequently forage in peripheral areas, retreating to trees to mate.

Breeding habitat: Shallow, peaty waters in fens, reedy ditches, marshes and bogs with lush vegetation and near woodland; sometimes slow-flowing drains and canals. Areas of open water are typically small and reduced by encroaching emergent vegetation.

LOOK-ALIKES (in range)
Other emeralds (*pp. 232–249*)
Eurasian Baskettail (*p. 250*)

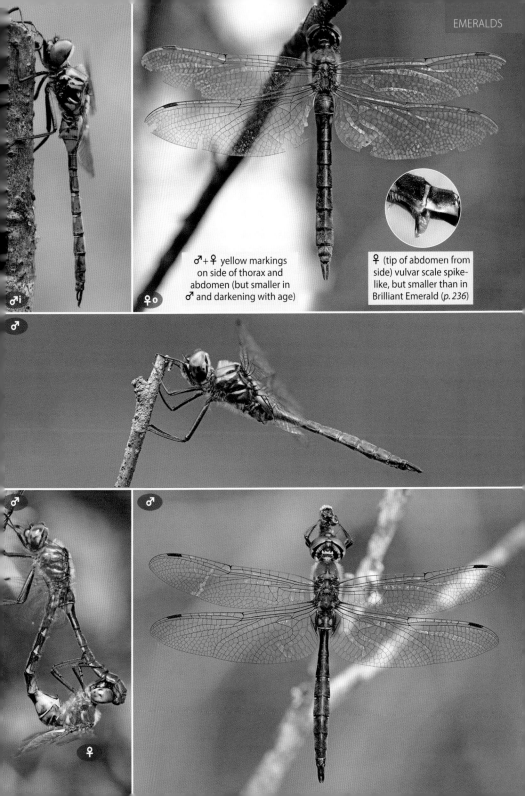

♂i

♀o

♂+♀ yellow markings
on side of thorax and
abdomen (but smaller in
♂ and darkening with age)

♀ (tip of abdomen from
side) vulvar scale spike-
like, but smaller than in
Brilliant Emerald (*p. 236*)

♂

♂

♀

♂

LC **Northern Emerald**

Somatochlora arctica ×1·5

Moorland Emerald

| Scarce and local ? |

| Peat bog |

J F M A M J J A S O N D

| **Overall length:** | 45–51 mm |
| **Hindwing:** | 28–35 mm |

♂ + ♀ yellow spot on each side of 'face'

A relatively small, dark, slender emerald, often found with the very similar Alpine Emerald (*p. 244*). Rather elusive and a specialist of moorland peat bogs.

Identification: Medium-sized, but one of the smaller emeralds; yellow spot on each side of 'face', next to **bright green** eyes; thorax metallic **bronzy-green**; abdomen much **darker**, almost black, base of S3 with incomplete yellow ring. ♂ Abdomen narrowly waisted at S3 (more so than on Alpine Emerald) and slightly swollen at S5–7; upper appendages large, **calliper-shaped** from above, with three irregular 'teeth' on underside; rear margin of hindwing is acutely angled. ♀ Abdomen has parallel sides, S3 with **two large yellow spots** on top, S9 with **long, blunt** vulvar scale that barely projects and reaches slightly beyond tip of S9; hindwing has rounded rear margin. VARIATION: Immatures have brown eyes.

Behaviour: Males fly low and erratically over bog pools, avoiding larger areas of open water. Rests and forages high among trees, but mates in lower shrubs. Eggs laid alone in flight, into open water over bog-mosses (*Sphagnum* spp.) or into wet peat.

Breeding habitat: Breeds in tiny areas of open water and shallow bog pools with abundant bog-mosses where few other dragonflies can survive; in moorland, tundra and lowland bogs, often close to woodland.

LOOK-ALIKES (in range)
Other emeralds (*pp. 232–249*)

♀ S3 with pair of large yellow spots

♂ + ♀ abdomen lacks obvious green

♀ vulvar scale long, flat, barely projecting, extends at least to S9

♂ abdomen distinctly waisted

♂ appendages calliper-shaped

LC Alpine Emerald

Somatochlora alpestris ×1·

Common ?

Bog ponds & lakes

J F M A M J J A S O N D

| Overall length: | 45–50mm |
| Hindwing: | 30–34mm |

A relatively small, dark emerald, found at small boggy pools in Alpine and Boreal regions, overlapping in range with Brilliant Emerald (*p. 236*), Northern Emerald (*p. 242*) and, in the far N, Treeline Emerald (*p. 246*).

Identification: Medium-sized; one of the smaller emeralds, but slightly more robust than Northern Emerald; yellow spot on each side of 'face', next to **bright green** eyes; thorax metallic **bronzy-green**; abdomen much **darker**, almost black, base of S3 with incomplete **whitish** ring, base of S4 with tiny whitish flecks; forewing usually has **two** cubito-anal cross-veins (one in other *Somatochlora* species). ♂ Abdomen narrowly waisted at S3 (but less so than in Northern Emerald) and slightly swollen at S5–7; upper appendages strongly **double-angled** on outer edges from above, with convergent tips and two small 'teeth' on underside; rear margin of hindwing is acutely angled. ♀ Abdomen shorter and thicker than in Northern Emerald with parallel sides, S9 with long, **triangular** vulvar scale projecting beneath; hindwing has rounded rear margin. VARIATION: Immatures have brown eyes.

Behaviour: Males patrol low and erratically over pools, chasing other emeralds. Both sexes may be found resting on vegetation and shrubs nearby. Adults are susceptible to summer snowfall, although larvae are tolerant of both freezing and desiccation.

Breeding habitat: Breeds in peaty ponds in bogs and peat workings with abundant bog-mosses (*Sphagnum* spp.). Less often found in acidic montane ponds and lakes set in meadows or woodland.

♂ + ♀ yellow spot on each side of 'face'.

LOOK-ALIKES (in range)
Other emeralds (*pp. 232–249*)

♀ (tip of abdomen from side) vulvar scale long, triangular, projecting

♂ + ♀ abdomen lacks obvious green; S3 with incomplete whitish ring

♂ forewing usually has two cubito-anal cross-veins (all other *Somatochlora* species have only one)

♂ appendages have double-angled outer edges

🄳🄳 Treeline Emerald

Somatochlora sahlbergi × 1·

Rare and local ❓

Bog ponds

J F M A M J J A S O N D

Overall length:	48–50 mm
Hindwing:	30–33 mm

♂ + ♀ yellow spot on each side of 'face'

A robust looking, dark emerald with the most northerly distribution of any dragonfly; found N of the Arctic Circle and overlapping in range with Brilliant Emerald (*p. 236*), Northern Emerald (*p. 242*) and Alpine Emerald (*p. 244*). One of the most difficult species to see in Europe due to its remote breeding areas.

Identification: Medium-sized, relatively robust emerald; yellow spot on each side of 'face', next to **bright green** eyes; thorax metallic **bronzy-green**; abdomen much **darker**, almost black, base of S3 with incomplete **whitish** ring; forewing has one cubito-anal cross-vein, like other *Somatochlora* emeralds (usually two in Alpine Emerald) – see *opposite*. ♂ Abdomen narrowly waisted at S3 (but less than in Northern Emerald and slightly swollen at S5–7; from above, upper appendages diverge slightly from base but **strongly bent** inwards and downwards near tip, with **no** 'teeth' on underside and ending in upcurved hooks; rear margin of hindwing is acutely angled. ♀ Abdomen shorter and thicker than in Northern Emerald with parallel sides, S9 with vulvar scale short and **barely projecting** beneath, tip notched from below (rounded in other emeralds in N); hindwing has rounded rear margin. VARIATION: Immatures have brown eyes; base of S4 may show tiny whitish flecks.

Behaviour: Rarely seen, spending much time perched in stunted vegetation and only flying in warm, calm conditions; unusually for an emerald, may sit horizontally on the ground. Numbers peak in late July.

Breeding habitat: Breeds in palsa (mires and bogs where peaty pools are created by melting ice lenses); these are fringed with bog-mosses (*Sphagnum* spp.) and sedges (*Carex* spp.) in the boreal zone (tundra and taiga), in areas of permafrost where habitat is in pristine condition.

LOOK-ALIKES (in range)

Brilliant Emerald (*p. 236*)
Northern Emerald (*p. 242*)
Alpine Emerald (*p. 244*)

♀

♂+♀ abdomen lacks obvious green; S3 with incomplete whitish ring

♀

♀ vulvar scale short, barely projecting

♂

♂

♂ forewing has one cubito-anal cross-vein (Alpine Emerald (p. 244) usually has two)

♂ appendages have sharply angled tips, ending in tiny hooks

LC Orange-spotted Emerald

Oxygastra curtisii × 1·5

Fairly common =

River, stream, canal, lake

J F M A M J J A S O N D

Overall length:	47–54 mm
Hindwing:	33–36 mm

♂ + ♀ 'face' all-dark

This relict species is the sole member of its genus and is no longer considered to be a member of the family Corduliidae. Its exact taxonomic position is uncertain (hence being placed as *incertae sedis*), but its metallic colours certainly justify it being called an emerald. It is almost endemic to SW Europe (known also from three sites in Morocco).

Identification: Medium-sized; longer and more slender than Downy Emerald (*p. 232*), which also flies with abdomen raised in an arc. 'Face' all-dark, eyes and thorax **metallic green**; abdomen slender, dark **bronzy-green** marked with narrow, **orange-yellow markings** along top of S1–7 and small but obvious **pale yellow spot on S10**. ♂ Abdomen **very thin, broader around S8**; wings amber at base with hind margin less angular than in other emeralds. ♀ Base and tip of abdomen only slightly broader than the rest; front of wings extensively, although variably, **suffused amber**. VARIATION: Immatures have pinkish-brown eyes and more extensive amber in wings.

Behaviour: Males fly rather erratically low over short, well-spaced territories along the margins of slow-flowing, partially shaded stretches of water. Hovers rarely (unlike Downy Emerald, which does so frequently). Usually found close to breeding waters, feeding around trees, often at canopy height, but sometimes rests on vegetation and shrubs nearby.

Breeding habitat: Associated with muddy sediment, typically slow-flowing, tree-lined rivers and streams, and sometimes canals and lakes.

LOOK-ALIKES (in range)
Downy Emerald (*p. 232*)
Brilliant Emerald (*p. 236*)
Yellow-spotted Emerald (*p. 240*)
Clubtails (*pp. 178–191*)
Pincertails (*pp. 194–203*)

♀ abdomen more extensively orange-yellow than ♂

♀ + IMMATURES wings often with extensive amber suffusion

♀ i

♀

♂

♂

♂ curved abdomen apparent in flight (see *p. 230*), resembling Downy Emerald (*p. 232*) and pincertails (*pp. 194–203*)

♂ very slender abdomen with slightly clubbed tip and pale 'tail light' on S10

BASKETTAIL & CHASERS

FAMILY | **Corduliidae & Libellulidae**

GENERA | *Epitheca & Libellula*

4 SPECIES

Medium–large dragonflies (mean lengths 44–60 mm) with **black at base of hindwing** crossed by yellow veins; two species mainly brownish with sexes alike; males of other two species have pruinose abdomen. Breed in standing or slow-flowing waters; eggs laid in flight, by female alone, flicking tip of abdomen into water; male may hover nearby.

Key features of Baskettail & chasers

◆ black at base of hindwings
◆ standing or slow-flowing waters

Identify species by pattern of black in wings | abdomen shape, colour and pattern | behaviour

LC Eurasian Baskettail

Epitheca bimaculata × 1:

Scarce and local =

Lake

NORTH
SOUTH
J F M A M J J A S O N D

Overall length:	55–65 mm
Hindwing:	36–44 mm

An elusive, persistent flyer related to emeralds, but lacks metallic hue and is patterned rather like Four-spotted Chaser (*p. 252*).

Identification: Large; thorax yellow-brown with black stripes; Abdomen long and black with **yellow-brown markings** on side, darker and joining on top of S1–2, decreasing in size to little or none on S9–10; eyes dull blue-green; legs very long and spidery; wings with **black marking** (slightly larger in female) at centre of hindwing base and variable **yellow tint** across front of all wings. ♂ Upper appendages close at base and curved out. ♀ Upper appendages apart at base, often held parallel. VARIATION: Immatures have brown eyes; abdomen brighter than mature adult, with more extensive ochre and irregular black pattern on top of S3–10, like immature Blue Chaser (*p. 256*); wings extensively suffused golden.

Behaviour: Emerges in May–June, leaving behind long-legged, spidery-looking exuviae (cast larval skins). Patrols fast for long periods over open water at knee height, keeping well away from edges. Gelatinous string of eggs held in 'basket' under tip of abdomen, expanding when dipped in flight by female alone into floating plants. Feeds around trees, where rests vertically, unlike Four-spotted Chaser.

Breeding habitat: Deep woodland lakes, often bordered by trees or reeds, usually with extensive floating and submerged plants.

LOOK-ALIKES (in range)
Four-spotted Chaser (*p. 252*)
♀ Blue Chaser (*p. 256*)
Yellow-spotted Emerald (*p. 240*)

♀i

Most easily found (as immature) soon after emergence.

♀i

♂ + ♀ eyes dull blue-green (IMMATURES brown)

♂ + ♀ yellow-brown markings along side of abdomen resemble those of Yellow-spotted Emerald (*p. 240*)

♂

♂

♂i

♂ + ♀ wings with yellow tint across front

♂ + ♀ black patch in base of hindwings

LC **Four-spotted Chaser** *Libellula quadrimaculata* ×1:

Various standing waters

J F M A M J J A S O N D

Overall length:	40–48 mm
Hindwing:	32–40 mm

A rather dull chaser with unique dark spots on the wings. One of the most common and widespread dragonflies in the world, also occurring across Asia and N America (where it is known as Four-spotted Skimmer). Scarcer in S Europe, where usually in mountains.

Identification: Medium-sized; sexes similar, having **brown** eyes and translucent brown thorax and abdomen with irregular polygon shapes visible inside; abdomen tapered, final third or so **black** on top with **yellow** sides to S4–8; wings have **yellow** base, large triangular **black patches** with yellow venation at base of hindwings (also in forewings in other chasers) and diagnostic **dark spots at nodes**. ♂ Appendages close together at base. ♀ Appendages well separated at base. VARIATION: Nodal spots not apparent on newly emerged adults; immature bright ochre, front of wings with extensive yellow suffusion and veins, all becoming dull with age. Form *praenubila* has more black at nodes and dark smudges near wingtips.

Behaviour: Territorial males aggressive, intercepting passing intruders or females from perches on emergent or marginal vegetation. Females often harassed and mated by succession of males. Large numbers may roost communally in reeds and may form huge migrating swarms.

Breeding habitat: Wide range of open, well-vegetated standing waters, largest populations at acidic sites; sometimes at slow-flowing and brackish waters.

LOOK-ALIKES (in range)
Other chasers (*pp. 254–257*)
Eurasian Baskettail (*p. 250*)

♂+♀ diagnostic dark spots halfway along each wing

♀ o

♀ form *praenubila*

♂+♀ eyes brown

♂ form *praenubila*

♂+♀ form *praenubila* has dark pigmentation near wingtips and around nodes

♂+♀ abdomen brown with black tip and yellow sides to S4–8

♂

LC Broad-bodied Chaser

Libellula depressa ×1:

Very common	=

Various standing waters

J F M A M J J A S O N D

Overall length:	39–48 mm
Hindwing:	33–37 mm

A familiar dragonfly that wanders freely in search of new ponds, sometimes appearing within hours of their creation! Like other dragonflies, climate change has enabled it to expand its range, for example moving 100 km N in Britain in the last 50 years.

Identification: Medium-sized; eyes and thorax brown, latter with **pale** shoulder stripes; abdomen short, **very broad** and flat, with **yellow** sides to S3–7; all wing bases with large **dark patches**. ♂Shoulder stripes pale bluish, darkening with age; abdomen dark at base and tip, S3–9 with **pale-blue**, eventually covering yellow edges, pruinescence abraded by female's legs during mating to produce dark patches; appendages relatively long and close together. ♀ Shoulder stripes whitish; abdomen **yellowish-brown**, even broader than that of male with larger yellow markings on side and dark tip; appendages relatively short and held apart. VARIATION: Abdomen is mainly bright ochre and patterned as female in immatures; old females may become pruinose to some extent.

Behaviour: Territorial males aggressive, perching on marginal vegetation between knee- and waist-height and intercepting passing intruders or females. Flight fast and erratic. Mating usually very brief and in flight, followed by egg-laying in flight by flicking tip of abdomen into water, usually with male hovering nearby; females sometimes visit water when males absent to avoid being harassed.

Breeding habitat: Favours small standing water bodies, such as ponds, small lakes and ditches; occasionally slow-flowing and brackish waters, but uncommon at acidic sites. Especially fond of waters with open, bare margins and often the first colonist of new ponds and flooded mineral workings.

LOOK-ALIKES (in range)
Other chasers (*pp. 252–257*)
♂ skimmers (*pp. 262–281*)

♀

♀ old individuals may become pruinose, appearing like dark ♂

♀ o

♂ + ♀ eyes brown

♂ + ♀ combination of dark wing bases and broad abdomen is unmistakable

♂

♂

♂ + ♀ S3–7 sides yellow

LC Blue Chaser

Libellula fulva ×1

Scarce Chaser

| Locally common | = |

River, stream, lake, canal

J F M A M J J A S O N D

| Overall length: | 42–45 mm |
| Hindwing: | 32–38 mm |

An early summer species with notably colourful immatures, best appreciated during concerted emergence in spring. Similar to some skimmers (*pp. 262–281*) but short, squat and usually perches on vegetation rather than bare ground.

Identification: Medium-sized, shorter and bulkier than similar skimmers; eyes **blue-grey**, thorax dark brown with **pale** shoulder stripes; abdomen **quite broad**, especially in female; wings with **small dark patches** at bases (smaller than other chasers), triangular on hindwing and reduced to streak on forewing, yellow tinge across front and often **dusky tips**, especially in females. ♂ Thorax blackish; abdomen with pruinose **pale blue** S3–7 and blackish base and **tip** (which is more restricted in Turkey); dark 'mating scars' (pruinescence scratched off by female's legs) halfway along abdomen often obvious. ♀ Thorax olive-brown; abdomen ochre-brown, with black line along top broadening towards tip into series of **bell-shapes**; wings have yellow suffusion across front. VARIATION: Immatures have brown eyes, rich brown thorax and **orange** abdomen, patterned as female; wings with amber suffusion and yellow veins across front. Female abdomen darkens with age, rarely becomes pruinose.

Behaviour: Territorial males aggressive, perching on marginal vegetation up to 1 m above water (not on ground as many skimmers) and intercepting passing intruders or females. Mating occurs while perched and lasts for 15–30 minutes, longer than other chasers.

Breeding habitat: Lowland neutral or base-rich slow-flowing small rivers and streams, and a range of standing waters. Breeding sites have dense beds of tall emergent and marginal vegetation growing in organic sediment.

LOOK-ALIKES (in range)
Eurasian Baskettail (*p. 250*)
Other chasers (*pp. 252–255*)
♂ skimmers (*pp. 262–279*)

256

♀i

♀

♂ + ♀ black 'bell shapes' on abdomen, most obvious on IMMATURES

♂ + ♀ eyes blue-grey (when mature)

♀o

♂ + ♀ many individuals have dark wingtips

♂ + ♀ black wing bases often hard to see against dark thorax

♂i

♂

♂ abdomen narrower than that of Broad-bodied Chaser (*p. 254*), but wider and shorter than those of Black-tailed Skimmer (*p. 262*) and White-tailed Skimmer (*p. 264*)

See also: Chasers *p. 250* | Darters *p. 294*

SKIMMERS

FAMILY | **Libellulidae**

GENUS | *Orthetrum*

10 SPECIES

Skimmers are small- to medium-sized 'perchers' (mean lengths 36–59 mm) inhabiting a range of still and flowing wetlands. The males of most species become pruinose pale blue on the top of the abdomen, which in some species may also extend onto the thorax. However, they differ from the similarly pruinose Blue Chaser (*p. 256*) and Broad-bodied Chaser (*p. 254*) in lacking dark wing bases. They often sit on the ground, rather than on vegetation like chasers. Males can be difficult to identify, with examination of the secondary genitalia sometimes being needed to confirm identification, although this is rarely necessary in Europe. The yellowish-brown females, however, are often more distinctive than males. Both sexes have antenodal cross-veins in the forewing that are yellow in all species except vagrant Desert Skimmer, although these fade with age. Often seen skimming low over water or perched on the ground or among low vegetation. Females lay eggs into water, alone or guarded by males.

Key features of skimmers

◆ body pruinose (males of most spp.)
◆ 10–14 yellow antenodal cross-veins (dark in vagrant Desert Skimmer)
◆ clear wings

Identify species by presence and shade of pruinescence (males) | colour of 'face' | presence of pale stripes on side of thorax and between wing bases | shape of abdomen | pattern of dark markings on abdomen (females) | colour and size of wing-spot | habitat | geographic location

Key identification features

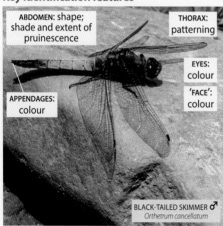

ABDOMEN: shape; shade and extent of pruinescence

THORAX: patterning

EYES: colour

'FACE': colour

APPENDAGES: colour

BLACK-TAILED SKIMMER ♂
Orthetrum cancellatum

WING: yellow antenodal cross-veins

WING: wing-spot colour and size

WING: costa

WING: radius

ABDOMEN: pattern of dark markings

KEELED SKIMMER ♀
Orthetrum coerulescens

Most skimmers have blue pruinescence (males and some females) and show 10–14 antenodal cross-veins, which are yellow in most species. In comparison, the slightly smaller darters (*pp. 300–321*) have 7–9 antenodal cross-veins, while dropwings (*pp. 324–331*) have 10–13. In darters and dropwings the antenodal vein nearest the node does not continue beyond the sub-costa (and is therefore denoted as ·5), whereas in skimmers it does (and is referred to as 'complete').

complete

Skimmer

incomplete

Darter

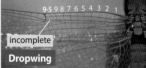

incomplete

Dropwing

Comparison of antenodal cross-veins

Skimmers compared (*Orthetrum* species)

Species	Male abdomen	Female abdomen	Thorax	Wing-spot	Other features	Range; main habitat
Black-tailed Skimmer *O. cancellatum* (p. 262)	Blue with black tip and appendages	2 parallel black lines and black tip	Relatively unmarked	Black	Dull 'face'	Widespread; open standing waters
White-tailed Skimmer *O. albistylum* (p. 264)	Pale blue with black tip; appendages usually white	2 curving black lines; white tip	2 indistinct pale stripes; pale between wing bases	Black	Pale blue 'face'	Southern; open standing waters
Keeled Skimmer *O. coerulescens* (p. 266)	Blue; relatively narrow	Thin black central line with short bars or spots to side	Pale shoulder stripes often obvious, but side relatively unmarked; pruinose in males in south	Orange, > width of abdomen	Dull 'face'	Widespread; vegetated slow-flowing waters
Southern Skimmer *O. brunneum* (p. 268)	Pale blue; relatively broad	Thin black central line with spots either side; edge of S8 bulges	Relatively unmarked; pruinose in males	Orange, < width of abdomen	Pale blue 'face'	Southern; open flowing waters
Yellow-veined Skimmer *O. nitidinerve* (p. 270)	Pale blue, often with narrow yellowish band on S1	Thin black central line with spots either side	Relatively unmarked; pruinose in males	Yellow-orange, relatively large	Outer part of costa yellow; radius yellow from wing base to node	Western half of Mediterranean – localized; arid waters
Desert Skimmer *O. ransonnetii* (p. 272)	Very pale blue with darker bands across (triangles in immatures)	Discontinuous central line with dark band across (triangles in immatures)	Pruinose in males	Orange	Antenodal cross-veins black or pruinose; eyes brown above	Vagrant from N Africa; often perch vertically on rocks to keep cool
Epaulet Skimmer *O. chrysostigma* (p. 274)	Blue; waisted with swollen base	Thin black central line with spots either side	White 'epaulet' stripe with dark borders on side (pruinose in mature males); females pale between wing bases	Orange	Small orange patch in base of hindwing	South-western and south-eastern; standing and flowing waters
Long Skimmer *O. trinacria* (p. 276)	Dark blue, paler at base; long and thin with swollen base	Thick black central line and tip; long and thin with swollen base and long appendages	Relatively unmarked	Pale, long	Much the longest skimmer (more than 50 mm)	South-western; standing waters
Small Skimmer *O. taeniolatum* (p. 278)	Blue thorax and abdomen	Thick black central line bordered yellowish, with brown sides	Side has 2 pale stripes with dark borders in front (pruinose in mature males); pale between wing bases	Orange, relatively short	Eyes brown above	South-eastern; standing and flowing waters
Slender Skimmer *O. sabina* (p. 280)	Black with pale side patches and tip; very thin with swollen base	Black with pale side patches and tip; thin with swollen base	Pale stripe on shoulders and side; pale between wing bases	Yellowish	Greenish eyes	South-eastern; standing and slow-flowing waters

Skimmers compared (breeding species) – males
These comparison plates show the nine breeding species of skimmer in Europe.

'FACE': dull

WING-SPOTS: black

S7/8—10: black

APPENDAGES: black

Black-tailed Skimmer
Orthetrum cancellatum (p.262)

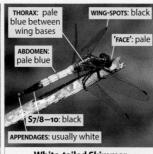

THORAX: pale blue between wing bases

WING-SPOTS: black

'FACE': pale

ABDOMEN: pale blue

S7/8—10: black

APPENDAGES: usually white

White-tailed Skimmer
Orthetrum albistylum (p.264)

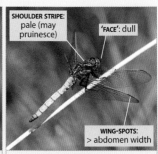

SHOULDER STRIPE: pale (may pruinesce)

'FACE': dull

WING-SPOTS: > abdomen width

Keeled Skimmer
Orthetrum coerulescens (p.266)

ABDOMEN: pale blue, quite broad

'FACE': pale blue

WING-SPOTS: < abdomen width

Southern Skimmer
Orthetrum brunneum (p.268)

COSTA: yellow on outerwing

RADIUS: yellow on innerwing;

Yellow-veined Skimmer
Orthetrum nitidinerve (p.270)

THORAX SIDE: pale stripe ('epaulet' (often pruinose))

ABDOMEN: waisted, swollen at base

Epaulet Skimmer
Orthetrum chrysostigma (p.274)

ABDOMEN: mainly dark; long, thin, swollen at base

WING-SPOTS: long, pale

Long Skimmer
Orthetrum trinacria (p.276)

THORAX SIDE: 2 pale stripes (often pruinose)

EYES: top brown

Small Skimmer
Orthetrum taeniolatum (p.278)

ABDOMEN: black with pale patches; thin with swollen base

APPENDAGES: pale

Slender Skimmer
Orthetrum sabina (p.280)

Other similar dragonflies

Pruinose male chasers and whitefaces may look similar to many skimmers but always have dark wing bases. Female darters (see *opposite*) have fewer (6·5–8·5), dark antenodal cross-veins (yellow in most skimmers).

Chasers (pruinose species)
Libellula (pp.254–257)

Whitefaces (pruinose species)
Leucorrhinia (pp.290–293)

Skimmers compared (breeding species) – females

WING-SPOTS: black

APPENDAGES: black

ABDOMEN: 'ladder' pattern

Black-tailed Skimmer
Orthetrum cancellatum (p. 262)

THORAX TOP: pale central stripe

ABDOMEN: 'ladder' pattern

S10: white

APPENDAGES: white   WING-SPOTS: black

White-tailed Skimmer
Orthetrum albistylum (p. 264)

S4–7: short dark bar across 'keel'

S8: slight bulge on edge

SHOULDER STRIPE: pale

Keeled Skimmer
Orthetrum coerulescens (p. 266)

S8: bulge on edge

S4–7: black dot either side of 'keel'

Southern Skimmer
Orthetrum brunneum (p. 268)

S8: no bulge on edge

RADIUS: yellow on innerwing;

COSTA: yellow on outerwing

Yellow-veined Skimmer
Orthetrum nitidinerve (p. 270)

THORAX TOP: pale stripe between wingbases

THORAX SIDE: pale 'epaulet' stripe

Epaulet Skimmer
Orthetrum chrysostigma (p. 274)

ABDOMEN: black with large pale patches; long, thin, swollen at base

WING-SPOTS: long, pale

APPENDAGES: long

Long Skimmer
Orthetrum trinacria (p. 276)

ABDOMEN: thick black stripe on top

THORAX SIDE: 2 pale stripes

Small Skimmer
Orthetrum taeniolatum (p. 278)

ABDOMEN: black with pale patches; thin with swollen base

THORAX TOP: pale stripe

APPENDAGES: pale

THORAX SIDE: pale stripe

Slender Skimmer
Orthetrum sabina (p. 280)

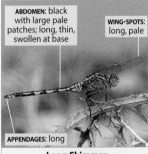

Darters
Sympetrum (pp. 298–321)

As with other dragonflies, some female skimmers become pruinose. Such females can usually be told from males by their relatively short, widely spaced appendages.

KEELED SKIMMER ♀
Sympetrum striolatum

LC Black-tailed Skimmer

Orthetrum cancellatum ×1·

Very common =

Various standing and slow-flowing waters

J F M A M J J A S O N D

| Overall length: | 44–50 mm |
| Hindwing: | 35–40 mm |

The commonest and most widespread of the skimmers, indeed one of Europe's commonest dragonflies, typically found resting on bare surfaces near water.

Identification: Relatively large skimmer; 'face' dull yellowish; thorax with thin, incomplete black shoulder stripes; abdomen relatively broad, tapering evenly from S2–3 to **black** appendages (white in otherwise very similar White-tailed Skimmer (*p. 264*)); wings clear with yellow costa and black wing-spots (superficially similar Blue Chaser (*p. 256*) has dark base to hindwing). ♂ Eyes **greenish-blue**; thorax side pale brown with two thin black lines, darker on top with indistinct black shoulder stripe; abdomen pruinose pale blue except for brown base and **black S7/8–10**.
♀ Eyes olive-brown; thorax side yellow with two thin black lines, darker on top with incomplete shoulder stripes; abdomen has two broad black lines running along top, forming ladder-like pattern with 'rungs' at segment junctions (lines on White-tailed Skimmer made up series of obvious black arcs). VARIATION: Immatures bright yellow, with orange arcs on side of abdomen and black lines on top that become pruinose on maturing males. Female becomes dull greyish with age, some pruinose.

Behaviour: Typically sits on bare surfaces, males frequently flying out over water, zigzagging fast and low.

Breeding habitat: Favours lowland lakes, large ponds, mineral and peat workings, canals and slow-flowing rivers with at least partially exposed margins. Tolerates high fish densities and brackish water, especially in the Baltic states. Readily colonizes new sites, especially while open ground persists.

LOOK-ALIKES (in range)
Other skimmers (*pp. 264–281*)
Blue Chaser (*p. 256*)

♂i

♂ + ♀ 'face' rather dull yellowish

♂ eyes greenish-blue (when mature)

♂

♂ abdomen longer and narrower than that of superficially similar Blue Chaser (*p. 256*), and lacks dark wing bases of that species

♂ S7/8–10 and appendages black

♂ + ♀ wing-spots black

♂ + ♀ appendages black

♀

♀

♀ eyes olive-brown

♀ + IMMATURES black 'ladder' pattern on abdomen

LC **White-tailed Skimmer**

Orthetrum albistylum ×1

Common △

Variety of standing and flowing waters

J F M A M J J A S O N D

Overall length:	45–50 mm
Hindwing:	33–38 mm

Sleek, pale skimmer found across much of C and E Europe. Similar to slightly darker Black-tailed Skimmer (*p. 262*), which is always found within the same general area and uses similar habitats. Spreading N as a result of warmer summers and habitat creation.

Identification: Relatively large skimmer; eyes green; 'face' **very pale**; thorax brown with thin, incomplete black shoulder stripes, **two broad pale stripes** on side and **pale stripe on top** extending from between wing bases to front of thorax; abdomen relatively narrow with **white appendages** (black in Black-tailed Skimmer); wing-spots black, larger than in Black-tailed Skimmer. ♂ 'Face' **pale blue**; stripes on thorax **bluish**; abdomen **slender**, pruinose **very pale blue** except for brown base and **black S7/8–10**.
♀ Base colour pale yellow; abdomen has two broad black lines running along top formed by a series of **arcs** (straight lines on Black-tailed Skimmer), creating ladder-like pattern with 'rungs' at segment junctions; **S10 white**, enhancing white appendages. VARIATION: Immatures patterned as female but whitish-yellow. Females becomes dull with age; some males have black appendages.

Behaviour: Sits on open ground like Black-tailed Skimmer, but often also on plant stems. Even more aggressive than Black-tailed Skimmer, sometimes eating other dragonflies.

Breeding habitat: A wide range of sunny ponds lakes and other natural waters, usually with lush floating and submerged plants; sometimes slow-flowing and intermittent streams in S, where larvae live in residual pools. Has spread N in recent decades, especially at artificial sites such as mineral workings and fishponds.

LOOK-ALIKES (in range)
Other skimmers (*pp. 262–281*)

♂

♂ + ♀ 'face' very pale, whitish

White 'tail' hard to see in ♂, easier in ♀ (beware confusion from occasional ♂ with black appendages, or pale clay deposit on tip of abdomen of ♀ Black-tailed Skimmer (*p. 262*)).

♂ S7/8–10 black; appendages usually white

♀i

♀

♂ + ♀ eyes green (when mature)

♂ + ♀ appendages usually white

♀

♂ + ♀ wing-spots black

♀ + IMMATURES pairs of black arcs down abdomen that form a 'ladder' pattern

♀ + IMMATURES S10 and appendages white

LC **Keeled Skimmer**

Orthetrum coerulescens ×1·5

Heathland Skimmer

| Common | = |
| Stream, bog, flush | |

J F M A M J J A S O N D

| **Overall length:** | 36–45 mm |
| **Hindwing:** | 28–33 mm |

The most common and widespread small skimmer of flowing waters. Populations in far SE Europe, ssp. *anceps*, were once considered to be a separate species (*O. ramburii*) on the basis of differences in secondary genitalia.

Identification: Relatively small skimmer; both sexes have dull brownish 'face'; dark brown top to thorax with **buff** shoulder stripes (fading with age in males); abdomen **slender** with thin dark line ('keel') along top; wings with **yellow costa** and **relatively large, orange** wing-spots (longer than width of abdomen). ♂ Eyes blue-grey; abdomen evenly tapering, **wholly powder-blue** apart from dark S1. ♀ Abdomen **ochre-yellow** with parallel sides and short **dark bars** crossing 'keel' close to segment divisions; bottom edge of S8 protrudes slightly; wing base with golden suffusion, which may extend out towards node. VARIATION: Immatures have extensive golden suffusion to wings. In male, pale shoulder stripes fade with age; thorax may become pruinose and 'face' bluish (especially in S, including ssp. *anceps* in far SE), resembling larger Southern Skimmer (*p. 268*); female abdomen darkens with age and may become pruinose.

Behaviour: Males have relatively small territories, defended from low perches in fast and erratic flight with frequent, brief spells of hovering before returning to the same spot. Wings often held forward at rest.

Breeding habitat: Streams, ditches and runnels, often rocky in S but also in acidic bogs, flushes and seepages in N.

LOOK-ALIKES (in range)
Other skimmers (*pp. 264–281*)
♀ darters (*pp. 300–321*)

♂

♂ abdomen entirely
powder-blue

♂ thorax typically brown, but
in the S may become pruinose

♂ + ♀ 'face' dull brownish

♂

♀

♂ + ♀ wing-spots orange, relatively
large (longer than abdomen width)

♀

♀ o

♀ abdomen yellow-ochre with
black 'keel' and short dark bars
close to segment divisions

LC Southern Skimmer

Orthetrum brunneum ×1·5

Common △

Various flowing and
standing waters

J F M A M J J A S O N D

| **Overall length:** | 41–49 mm |
| **Hindwing:** | 33–37 mm |

A rather robust 'blue' skimmer, very similar to Keeled Skimmer (*p. 266*) and easily overlooked, but larger and males more extensively pale blue. It is less common in the N of its range, but has increased there substantially since the 1990s.

Identification: Medium-sized; abdomen relatively **broad**; wing-spots **reddish-brown**, shorter than width of abdomen (longer than width of abdomen in Keeled Skimmer); centre of outerwing has row of usually **4–9 divided cells** (a feature also shown by Yellow-veined Skimmer (*p. 270*) and some Keeled Skimmers, although in that species the cells are usually undivided). Both sexes generally paler overall than Keeled Skimmer. ♂ 'Face' **very pale blue**; eyes bluish; pale powder-blue pruinescence covers **whole** of thorax and top of abdomen (Keeled Skimmer has dark S1 and thorax may become pruinose in S, but darker background colour and pale shoulder stripes may be visible). ♀ 'Face' pale; shoulder stripes indistinct (usually less conspicuous than on Keeled Skimmer); abdomen brownish-ochre with usually **discrete** small black spots on each segment, either side of fine central 'keel' line; bottom edge of S8 protrudes as small flange (unlike Yellow-veined Skimmer).

Behaviour: Males are territorial, often perching on bare ground, stones or other low perches: behaviour more typical of Black-tailed Skimmer (*p. 262*) than Keeled Skimmer.

Breeding habitat: A wide range of slow-flowing and still waters, including small flushes, streams and gravel pits, amid sparse vegetation and bare ground. Shows a preference for open, often new, habitats (as does Black-tailed Skimmer).

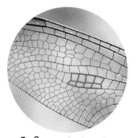

♂+♀ row of at least four divided cells in centre of outerwing (*highlighted in red*) (best confirmed from photo against a plain background)

LOOK-ALIKES (in range)
Other skimmers (*pp. 264–281*)
♀ darters (*pp. 300–321*)

♂

♂ abdomen, thorax and 'face' pale blue

♂+♀ 'face' pale

♂+♀ wing-spots reddish-brown (shorter than abdomen width)

♂

♀

♀ S8 bottom edge with small flange

♀

♂+♀ row of divided cells in centre of outerwing (see *opposite*) (as Yellow-veined Skimmer (*p. 270*))

♀ abdomen brownish-ochre with black 'keel' and pair of black spots close to segment divisions

Skimmers compared pp. 259–261

VU Yellow-veined Skimmer

Orthetrum nitidinerve ×1·5

Scarce and local ▽

Mainly small flowing waters

J F M A M J J A S O N D

Overall length:	46–50 mm
Hindwing:	31–38 mm

A relatively robust 'blue' skimmer, very similar to Southern Skimmer (*p. 268*), from which it is distinguished by bright yellow veins in the front of each wing. Scarce within its dry W Mediterranean range, where most often found at springs, seepages and intermittent streams.

Identification: Medium-sized; radius of innerwing and costa on outerwing both obviously **yellow** (see *opposite*); centre of outerwing usually has **row of divided cells** (as Southern Skimmer); **large yellow-orange** wing-spots **longer** than width of abdomen (as in Keeled Skimmer (*p. 266*), but unlike Southern Skimmer in which wing-spots are shorter than width of abdomen); abdomen relatively broad and parallel-sided (but **slightly** narrower than in Southern Skimmer). ♂ 'Face' **very pale blue**; eyes blue-grey; **very pale** powder-blue pruinescence covers **whole** of thorax and top of abdomen except for narrow yellowish band on S1 (Southern Skimmer slightly darker; Keeled Skimmer darker still). ♀ General appearance very plain (as Southern Skimmer); 'face' pale; lower edge of abdomen **straight**, **S8 not protruding** (no flange, unlike Southern Skimmer).

Behaviour: Territorial males typically perch on short stems beside watercourses.

Breeding habitat: Grassy springs, seepages and residual pools in small streams with intermittent flow, in arid areas; sometimes in larger watercourses and brackish waters.

♂+♀ row of at least four divided cells in centre of outerwing (*highlighted in red*) (best confirmed from photo against a plain background)

LOOK-ALIKES (in range)
Other skimmers (*pp. 264–281*)
♀ darters (*pp. 300–321*)

♂

radius   node   costa

♂+♀ distinctive pattern of
bright yellow veins in wing:
radius on innerwing and costa on
outerwing, with yellow 'zigzag' at
node (as indicated *above*)

♂ blue pruinescence
paler than other skimmers

♂+♀ wing-spots yellow-
orange, relatively large
(longer than abdomen width)

♂

♀

♀ S8 bottom edge
lacks flange

♀

♂+♀ row of divided cells
in centre of outerwing
(see *opposite*) (as Southern
Skimmer (*p. 268*))

Desert Skimmer

Orthetrum ransonnetii × 1·5

Ransonnet's Skimmer

| Vagrant |
| Desert oases |

J F M A M J J A S O N D

| Overall length: | 45–59 mm |
| Hindwing: | 27–32 mm |

As its name suggests, this is a species of desert regions of N Africa, the Middle East and S Asia. The sole European record to date came from Fuerteventura, Canary Islands, on 16 February 2018. Together with a faintly banded abdominal pattern, it can be identified by having dark and often pruinose, rather than yellow, antenodal cross-veins.

Identification: Medium-sized, with both sexes having top of eyes **brown** (similar to Small Skimmer (*p. 278*)) and small, orange wing-spots; however, abdomen has **obscure dark bands** becoming wider towards tips of segments, and wings have diagnostic **black or pruinose dark blue antenodal cross-veins** (yellow in all other European skimmers, although darker with age). ♂ Body extensively pruinose blue; tips of abdominal segments often with slightly darker blue 'shadow' of the triangle present in immatures. ♀ Pale brown; abdomen may show slightly darker triangles based on segment divisions; thin, discontinuous line along top of abdomen (continuous, thick, dark line on Small Skimmer, which also has two pale stripes across side of thorax). VARIATION: Immatures have **brown triangles at base of abdominal segments**, fading with age in both sexes.

Behaviour: Adults avoid overheating by perching on vertical rock surfaces to minimize the area exposed to direct sunlight. Females are rarely seen at breeding sites, where mating activity peaks around 09:30–10:00 am, avoiding the heat of the day.

Breeding habitat: Desert wadis and oases and surrounding rocky areas; in Morocco, known from the central High Atlas Mountains, where tolerates highly saline waters.

LOOK-ALIKES (in range)
Other skimmers (*pp. 264–281*)
♀ darters (*pp. 300–321*)

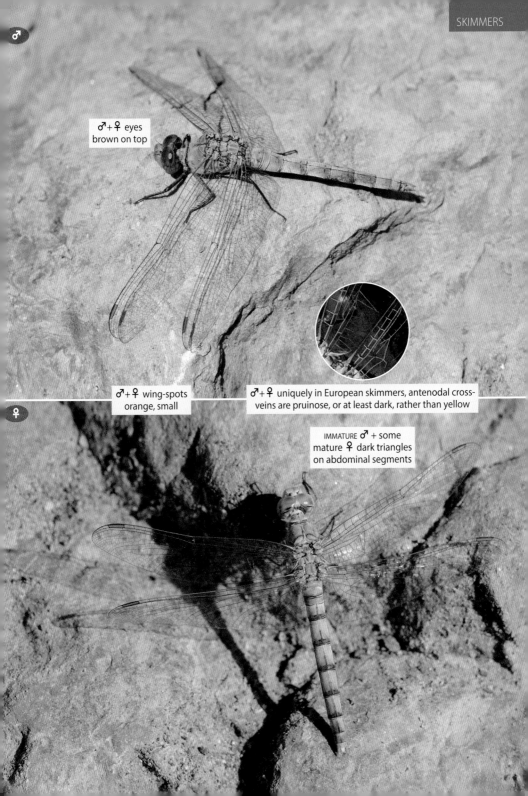

♂

♀+♂ eyes
brown on top

♂+♀ wing-spots
orange, small

♂+♀ uniquely in European skimmers, antenodal cross-
veins are pruinose, or at least dark, rather than yellow

IMMATURE ♂ + some
mature ♀ dark triangles
on abdominal segments

♀

LC Epaulet Skimmer

Orthetrum chrysostigma ×1·

Common =

Various flowing and
standing waters

J F M A M J J A S O N D

Overall length:	39–48 mm
Hindwing:	27–32 mm

A very common species in Africa that occurs sparingly in SE Europe, Malta and more widely in Iberia, where it has spread N in recent years. As with other skimmers, females and immatures are more easily identified than males, in this case having a distinctive single pale 'epaulet' on the side of the thorax.

Identification: Rather small, between Keeled Skimmer (*p. 266*) and Southern Skimmer (*p. 268*) in size; wings usually with small **yellow patch** at base; wing-spots orange. ♂ Pale blue pruinescence covers much of thorax and whole of abdomen, the latter distinctly **waisted** from above at S3–4 and with **swollen S1–2**; often shows **pale stripe** across front of side of thorax ('epaulet'), but faint when pruinose; eyes blue-green on top. ♀ Thorax with **conspicuous white stripe** outlined in black across front of side ('epaulet'), faint pale shoulder stripes, bordered on outside by black stripe, and **white stripe** between wing bases; abdomen brownish-ochre with usually **discrete** small black spots on each segment; eyes blue-brown. VARIATION: Immatures boldly marked with white stripes on thorax as female, abdomen with more-or-less dark stripes down either side of thin dark mid-line; females may have pinkish rear side of thorax.

Behaviour: Males patrol low over water before returning to regular perch on stem, rock or ground.

Breeding habitat: Sunny, sparsely vegetated flowing and standing waters in dry, often rocky, lowland areas. These include seasonal streams, large rivers and both natural and artificial ponds and lakes.

LOOK-ALIKES (in range)
Other skimmers (*pp. 264–281*)
♀ darters (*pp. 300–321*)

♂ may show a hint of pale stripe ('epaulet') on side of thorax

♂ eyes blue-green on top

♂ abdomen distinctly waisted

♂ + ♀ wings usually with small yellow patch at base

♀ + IMMATURES white stripe on top of thorax between wing bases

♀ + IMMATURES conspicuous white stripe ('epaulet') on side of thorax

LC Long Skimmer

Orthetrum trinacria ×1·5

Scarce and local =	
Lake, pond	

J F M A M J J A S O N D

Overall length:	51–67 mm
Hindwing:	34–38 mm

This common Africa species is by far the largest skimmer in Europe, where it has only been known since 1972. Northward spread from Africa has been assisted by climate change and the creation of reservoirs in Iberia, although to date few have been seen in Turkey and adjacent Greek islands.

Identification: Medium-sized, although very long for a skimmer, even recalling a small hawker (*pp. 136–139*) in flight; in both sexes abdomen **long and slender**, with **swollen S1–2** (from side) and more extensive black markings above than other skimmers; appendages equally **long** in both sexes; wing-spots long and pale. ♂ Body thinly pruinose, dark blue-grey, abdomen **often with extensive black** areas; eyes blue. ♀ Body pale yellow; thorax with thin black lines on top and side; abdomen with thick black mid-line and discontinuous black edges leaving isolated pale patches; S8–10 mostly black; appendages 3× length of S10. **VARIATION:** Old females may become pruinose, appearing darker than in other skimmers.

Behaviour: Territorial, typically perching on vegetation over water. A strong flier and very aggressive, taking prey including butterflies and other dragonflies.

Breeding habitat: Lowland lakes, reservoirs, large ponds and marshes, typically with lush marginal vegetation.

LOOK-ALIKES (in range)
Other skimmers (*pp. 264–281*)

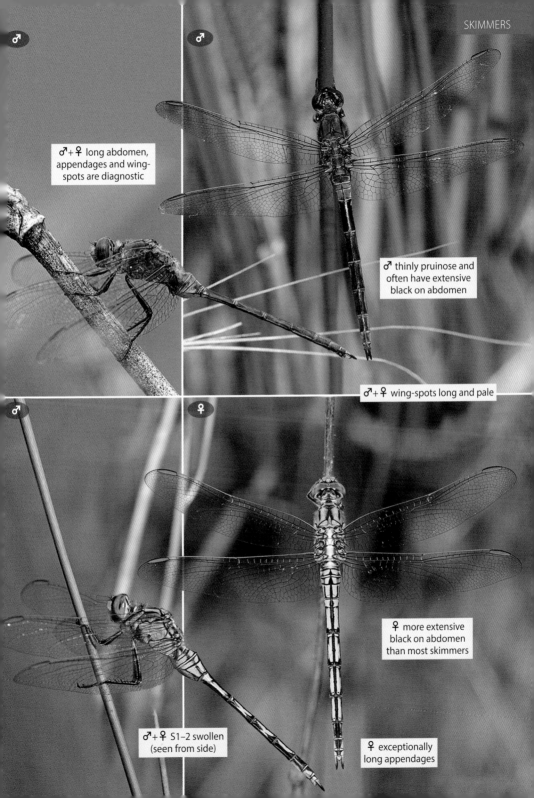

♂

♂

♂ + ♀ long abdomen, appendages and wing-spots are diagnostic

♂ thinly pruinose and often have extensive black on abdomen

♂ + ♀ wing-spots long and pale

♂

♀

♀ more extensive black on abdomen than most skimmers

♂ + ♀ S1–2 swollen (seen from side)

♀ exceptionally long appendages

LC Small Skimmer

Orthetrum taeniolatum × 1·5

Locally common =

River, lake

J F M A M J J A S O N D

| Overall length: | 33–38 mm |
| Hindwing: | 25–28 mm |

A tiny skimmer whose SW Asian range just reaches the far SE of the region. As with other skimmers, females are more easily identified than males, having a stripy appearance, especially on the thorax.

Identification: Small, with small orange-yellow wing-spots and **brown eyes**. ♂ Body entirely pruinose pale blue, although pale stripes on thorax may remain visible, especially between wing bases; abdomen tapers gradually from base to tip. ♀ Overall appearance well patterned and browner than other skimmers, with **boldly striped thorax** and abdomen; thorax with whitish stripes on shoulders, whitish stripe between wing bases and **two** whitish stripes across each side, each stripe bordered by black on one side; abdomen tricoloured, with **thick black central line** bordered by dull yellowish and then brown stripes down sides. VARIATION: Immatures more richly coloured than females, which become dull with age and may become pruinose.

Behaviour: Individuals, including territorial males, often sit on the ground, especially on rocks or pebbles.

Breeding habitat: Mainly slow-flowing rivers with expanses of sand or gravel; sometimes shallow standing waters, including dammed watercourses.

LOOK-ALIKES (in range)
Other skimmers (*pp. 264–281*)
♀ darters (*pp. 300–321*)

♂

♂

♂ body entirely pale blue (when mature)

♂ + ♀ eyes brown

♀

♀

♀ + IMMATURES stripy appearance, with whitish stripes on thorax and tricoloured abdomen

♀ + IMMATURES thick black line along top of abdomen

♀ o

♂

♀

LC Slender Skimmer

Orthetrum sabina ×1·

Green Skimmer, Sombre Skimmer, Green Marsh Hawk

| Locally common | = |
| Standing and flowing waters | |

J F M A M J J A S O N D

| Overall length: | 43–50 mm |
| Hindwing: | 28–33 mm |

An abundant Asian species whose range extends to Australia, N and NE Africa, the Middle East, SW Turkey and adjacent islands. Unlike other skimmers it does not become pruinose, but retains a distinctive, well-patterned appearance not unlike a clubtail (*pp. 178–191*).

Identification: Relatively large skimmer; sexes similar; thorax and base of abdomen **greenish-yellow** liberally striped with black, darkening with age to mostly brown with yellow stripes on shoulder and side; abdomen **slender** with **swollen** base and tip in side view, mostly black with pairs of long whitish patches on sides of S4–6; appendages **white** or mostly white; wing-spots pale. ♂ Does not become pruinose, as other skimmers; eyes green. ♀ Eyes green-brown. VARIATION: Extent of black variable; pale patches on abdomen reduce in size and darken with age.

Behaviour: Perches on twigs or the ground. Very active and wary, and somewhat nomadic.

Breeding habitat: A wide range of sunny lowland standing and slow-flowing waters; can tolerate brackish conditions.

LOOK-ALIKES (in range)
♀ other skimmers
(*pp. 264–279*)
Clubtails (*pp. 178–191*)

♂

♂ eyes green

♂ + ♀ overall appearance similar to a clubtail (*pp. 178–191*), but eyes meet

♂ + ♀ appendages pale

♀

♂

♂ + ♀ abdomen distinctive shape from side, with bulging base and tip

♂ + ♀ wing-spots pale

♀

WHITEFACES

GENUS | *Leucorrhinia*

FAMILY | **Libellulidae**

5 SPECIES

Small 'perchers' (mean lengths 34–36 mm), with a conspicuous white 'face' and dark spots at base of hindwing. The bodies of females and immatures are black with yellow markings that turn either red in the males of three species, or pruinose blue in the others. Immatures are difficult to tell apart. White veins extend outwards from the small wing-spots. Occur mainly in NE Europe and become very localized farther west and south. They emerge in spring and have a relatively short flight season.

Key features of whitefaces

◆ white frons ('face')
◆ dark hindwing base
◆ white 'rayed' veins beyond wing-spot
◆ dark thorax

Identify species by extent of red, yellow or pruinescence on abdomen

YELLOW-SPOTTED WHITEFACE
Leucorrhinia pectoralis

All whitefaces have a white frons and white veins radiating out from small, dark wing-spot.

SMALL WHITEFACE ♂
Leucorrhinia dubia

LILYPAD WHITEFACE ♂
Leucorrhinia caudalis

Three of the five whiteface species have black-and-red males, the other two have a partially pruinose abdomen; all have a dark thorax and wing bases.

Whitefaces compared – males

COSTA: basal half dark; outer half yellow

ABDOMEN: narrow; red markings narrow

Small Whiteface
Leucorrhinia dubia
(p. 284)

COSTA: all-yellow

ABDOMEN: fairly narrow; red markings fairly narrow

Ruby Whiteface
Leucorrhinia rubicunda
(p. 286)

ABDOMEN: broad; pale markings broad

S7: large yellow spot

Yellow-spotted Whiteface
Leucorrhinia pectoralis
(p. 288)

Black Darter *Sympetrum danae*
[for comparison]
(p. 300)

WING-SPOTS: black

S3–4: pruinose

APPENDAGES: white

Dark Whiteface
Leucorrhinia albifrons
(p. 290)

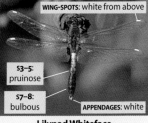

WING-SPOTS: white from above

S3–5: pruinose

S7–8: bulbous

APPENDAGES: white

Lilypad Whiteface
Leucorrhinia caudalis
(p. 292)

Whitefaces compared – females

COSTA: basal half dark; outer half yellow

ABDOMEN: narrow; yellow markings narrow

Small Whiteface
Leucorrhinia dubia
(p. 284)

COSTA: all-yellow

ABDOMEN: narrow; yellow markings wide

Ruby Whiteface
Leucorrhinia rubicunda
(p. 286)

ABDOMEN: broad; yellow markings broad

S7: large yellow spot

Yellow-spotted Whiteface
Leucorrhinia pectoralis
(p. 288)

Black Darter *Sympetrum danae*
[for comparison]
(p. 300)

PALE SHOULDER STRIPE: short

ABDOMEN: yellow markings small

APPENDAGES: white

Dark Whiteface
Leucorrhinia albifrons
(p. 290)

PALE SHOULDER STRIPE: long

ABDOMEN: yellow markings broad

S7–8: bulbous

APPENDAGES: white

Lilypad Whiteface
Leucorrhinia caudalis
(p. 292)

LC Small Whiteface

Leucorrhinia dubia ×1·5

White-faced Darter

Common =

Bog, pond, lake

J F M A M J J A S O N D

Overall length:	31–36 mm
Hindwing:	23–28 mm

This little dragonfly is the most widespread of the whitefaces and the smallest of the three species with black-and-red males. Like other species of northern and montane wetlands, it is rarer in southern locations.

Identification: Small. Both sexes have creamy-white frons, dark eyes, mostly black body, black legs, small black patch on wing base, and dark brown wing-spots with white veins radiating out; viewed from in front, costa **dark** from base to node, yellow from node to tip. ♂ Thorax has red shoulder stripes, irregular red blotches on side and red spots around wing bases; abdomen **narrow**, slightly waisted and black with red markings on top of S2–7, relatively narrow on S5–7 (broader on Ruby Whiteface (*p. 286*) and Yellow-spotted Whiteface (*p. 288*)). ♀ Patterned like male, but in yellow, which is more extensive on S1–7. **VARIATION:** Immatures black and yellow, as females; colour of female darkens with age to ochre or even red; as male ages, red areas on abdomen become darker and smaller and may be absent from S4–6.

Behaviour: Often perches on the ground, basking on pale ground or wood. Males less territorial than other darters, defending relatively small territories. Flight low, erratic and 'bouncy'. Tenerals disperse to nearby scrub or woodland, where mature adults roost.

Breeding habitat: Acidic bog pools with abundant bog-mosses (*Sphagnum* spp.), rushes (*Juncus* spp.) and sedges (*Carex* spp.), usually within woodland; such waters often result from small-scale peat digging and are usually fish-free. Able to breed in waterlogged bog-moss depressions without open water. In the N also found at lakes.

LOOK-ALIKES (in range)
Other whitefaces (*pp. 286–293*)
Black Darter (*p. 300*)

♂

♂ abdomen narrow

♂+♀ all whitefaces have white 'faces' and black wing bases

♂

♀ o

♀ with abdomen swollen with eggs may look bulky

♂+♀ basal half of costa dark, outer half yellow (clearest when viewed from in front)

♀

♀ coloured areas on S4–7 narrower than on other whitefaces

♂

♀

LC Ruby Whiteface

Leucorrhinia rubicunda × 1·

Northern White-faced Darter

Common ▽

Bog, lake, pond

NORTH
SOUTH
J F M A M J J A S O N D

Overall length:	31–38 mm
Hindwing:	27–31 mm

Principally a boreal species, but less widespread than Small Whiteface (*p. 284*), especially in the S, where it has declined and is threatened by climate change (fewer than 20 populations now persist in the N Alps). Slightly more robust than Small Whiteface, but with more extensive coloration on a broader abdomen, closely resembling Yellow-spotted Whiteface (*p. 288*).

Identification: Small, although slightly larger than very similar Small Whiteface, both species having a creamy-white frons, dark eyes, black legs, mostly black body with pale shoulder stripes and irregular blotches on thorax side and markings on S2–7. However, both sexes have **broader** coloured spots on abdomen than Small Whiteface (closer in size to those of Yellow-spotted Whiteface), **all-yellow** costa (black basal half in Small Whiteface), smaller dark patch on wing base, and reddish-brown wing-spots with white veins radiating out towards wingtip. ♂ Pale markings are red. ♀ Patterned like male, but in yellow, with spots on abdomen larger; very similar to Yellow-spotted Whiteface, but yellow spot on S2 smaller. VARIATION: Yellow suffusion on wings more extensive on immatures; markings on S2–6 darken with age, but spot on S7 remains distinct in male; markings on female may turn reddish.

Behaviour: Like Small Whiteface, perches quite close to the ground and basks in bare places, such as on tree stumps. Flight period earlier than other whitefaces.

Breeding habitat: Similar to Small Whiteface: acidic bogs, lakes and ponds with bog-mosses (*Sphagnum* spp.), often in woodland; but can tolerate at least moderately high fish densities.

LOOK-ALIKES (in range)
Other whitefaces (*pp. 284–293*)
Black Darter (*p. 300*)

♂ + ♀ whole of costa yellow (clearest when viewed from in front)

♂ abdomen slightly broader than that of Small Whiteface (*p. 284*), with more extensive coloured markings

♂

♂

♀ o

♂

♀

♀ as with other dragonflies, may take on ♂ coloration as matures

♀ i

LC Yellow-spotted Whiteface

Leucorrhinia pectoralis ×1·

Large White-faced Darter

Uncommon ▽

Bog, lake, pond

NORTH
SOUTH
J F M A M J J A S O N D

Overall length:	32–39 mm
Hindwing:	23–33 mm

The largest European whiteface, similar to Small Whiteface (*p. 284*) and Ruby Whiteface (*p. 286*), but with more extensive markings on a broader abdomen. Found in a wider variety of habitats than other whitefaces and relatively tolerant of warm conditions, although still scarce in the S of its range.

Identification: Small, although larger and more robust than similar Small and Ruby Whitefaces, all three species having a creamy-white frons, dark eyes, black legs, mostly black body with pale shoulder stripes, irregular blotches on thorax side and markings on S2–7. However, both sexes have **broader** abdomen with **broader** coloured spots than the other species, yellow costa darkening at base (black basal half in Small Whiteface), small blackish patch at wing base (smaller than in Small Whiteface), and dark wing-spots (red-brown in Ruby Whiteface) with white veins radiating out towards wingtip. ♂ Pale markings are dull reddish-brown, except for diagnostic **pale yellow spot on S7**; abdomen swollen around S6–7. ♀ Patterned like male, but in yellow, with spots on abdomen very wide, that on S2 forming **band across top** (rather than discrete spot, as in Ruby Whiteface) and that on S7 often slightly brighter than others; wing bases suffused yellow. VARIATION: Yellow suffusion on wings more obvious on immatures; markings on S2–6 darken with age and may turn reddish in female; spot on S7 remains pale in male and often so in female.

Behaviour: Territorial males perch on marginal vegetation, from which they fly to patrol through tall vegetation and over water. Occasional large-scale migrations have aided colonization, following declines in late 20th century.

Breeding habitat: Boggy pools and lakes, fens, ditches and even slow-flowing rivers and canals, typically with lush and diverse submerged and emergent vegetation, often peaty and mildly acidic. Can tolerate some nutrient enrichment and the presence of fishes, provided there is emergent vegetation in which larvae can shelter.

LOOK-ALIKES (in range)
Other whitefaces (*pp. 284–293*)
Black Darter (*p. 300*)

♂

♂ bright yellow spot on S7

♂+♀ costa yellow, darkening at base (clearest when viewed from in front)

♂

♂+♀ abdomen broader than in Small Whiteface (*p. 284*) and Ruby Whiteface (*p. 286*), with more extensive coloured markings

♀

♀ o

♀

♀ pale spot on S2 very large

LC Dark Whiteface

Leucorrhinia albifrons ×1·

Eastern White-faced Darter

Scarce and local =

Lake, bog

J F M A M J J A S O N D

Overall length:	33–39 mm
Hindwing:	23–28 mm

This species and Lilypad Whiteface (*p. 292*) differ from other whitefaces in having dark males with a pruinose base to an otherwise dark abdomen, lacking any red markings. It declined markedly in the 20th century due to loss of peat bogs and nutrient enrichment and continues to be threatened in the S and W of its range.

Identification: Small. Like all whitefaces, both sexes have a creamy-white frons, mostly black body, dark eyes, black legs and dark wing-spots with white veins radiating out; like Lilypad Whiteface, differs from other whitefaces in having **white appendages and dark base only to hindwing**; diagnostic **white sides to mouthparts**. ♂ Body dark, except for pale blue-grey pruinescence on S3–4 (less obvious than in Lilypad Whiteface); abdomen rather slender and slightly waisted (thicker and more 'clubbed' in Lilypad Whiteface). ♀ Thorax has **very short** yellow shoulder stripes, irregular yellow blotches on side and three yellow marks between wing bases; abdomen with **small** yellow markings on top of S2–6, **very narrow** on S4–6/7, with **yellow bands** across bases of S3–4; yellow suffusion in wing bases. VARIATION: Immatures black and yellow, as female; male pruinescence may extend thinly to thorax and S5, with age may appear very grey; dark females may have pale marks on only S2–3; yellow suffusion in female wings may be quite extensive.

Behaviour: Often perches in waterside trees and bushes, rarely on lilypads.

Breeding habitat: Favours lowland peat bogs and shallow woodland lakes with abundant floating and emergent plants; sites either acidic and fish-free or with vegetation in which larvae can seek refuge from fishes. Occurs up to 1400 m in Jura and Alps.

LOOK-ALIKES (in range)
Other whitefaces (*pp. 284–293*)
Black Darter (*p. 300*)

♂ abdomen rather slender with typically only S3–4 pruinose

♀ i

♂ + ♀ appendages white

♀ + IMMATURES yellow markings less extensive than other whitefaces

♂ + ♀ diagnostic white sides to mouthparts

LC Lilypad Whiteface

Leucorrhinia caudalis ×1·5

Dainty White-faced Darter

Scarce and local =

Lake, bog

NORTH
SOUTH
J F M A M J J A S O N D

| Overall length: | 33–37 mm |
| Hindwing: | 29–32 mm |

This species and Dark Whiteface (*p. 290*) are the two European whitefaces in which the males are dark with a pruinose base to an otherwise dark abdomen; they lack any red markings. Well named after the males' habit of perching on lilypads, although can occur where few or no floating leaves are present.

Identification: Small, appearing short, club-tailed and broad-winged. Like all whitefaces, both sexes have a creamy-white frons, dark eyes and legs, mostly black body, and white veins radiating out from wing-spots; like Dark Whiteface, differs from other whitefaces in having **white** appendages and dark base only to hindwing; distinguished by obviously **swollen** S7–8. ♂ Body dark, except for very pale blue-grey pruinescence on S3–5 (more obvious than in Dark Whiteface); wing-spots **white** on top (dark from below). ♀ Thorax has **complete** orange-yellow shoulder stripes, irregular yellow blotches on side and yellow marks between wing bases (larger than in Dark Whiteface); abdomen with yellow markings on top of S2–6, broader than in Dark Whiteface; yellow suffusion to wing bases; wing-spots dark. VARIATION: Immatures black-and-yellow, as female; some females have ill-defined dark patch behind each wing-spot.

Behaviour: Territories typically based on lilypads or other floating leaves, where males sit, often with abdomen raised, between flights out over water; sometimes perches on waterside stems.

Breeding habitat: Lowland lakes and ponds, including artificial waters, with deep, clear water, abundant submerged and often floating vegetation. Tolerates the presence of fishes better than other whitefaces. Rare and threatened in S of range following decline in the 20th century, although there has been some recent recovery in the W.

LOOK-ALIKES (in range)
Other whitefaces (*pp. 284–291*)
Black Darter (*p. 300*)

♀ complete yellow shoulder stripes

♂ white wing-spots (when seen from above)

♂+♀ distinctive club-shaped abdomen

♂+♀ mouthparts dark

♂ S3–5 pruinose

♂+♀ appendages white

DARTERS, SCARLET, DROPWINGS, PENNANT, PERCHER & GROUNDLING

FAMILY | **Libellulidae**

GENERA | *Sympetrum, Crocothemis, Trithemis, Selysiothemis, Diplacodes & Brachythemis*

19 SPECIES

Small (mean lengths 30–46mm), rather dainty red, orange, violet-pink or black (mature male) or yellowish-brown (female or immature) 'perchers', lacking dark bases to hindwings (but may be tinted amber). Typically sit on bare surfaces or prominent perches, from which they fly up and often return to same position. Eggs often laid in tandem. Can be migratory, or at least dispersive, turning up far from breeding sites and helping northward range expansions in recent decades. Most species fly in late summer and autumn, others from early spring to early winter. Of the 19 European species, five are common and widespread over large parts of Europe, four are found mainly in C and E Europe and the remaining ten are found in S Europe. This introductory section highlights the key features of each of the six 'sub-types'.

Key features of darters, scarlet, dropwings, pennant, percher & groundling

- small 'perchers'
- males red, orange, violet-pink or black
- amber or clear wing bases
- 6–12·5 antenodal cross-veins

Identify 'sub-types' by number of antenodal cross-veins on forewing | colour of abdomen

Other 'small' dragonflies with similar features

Unlike most of the slightly larger skimmers (*pp. 262–281*), the species covered here lack blue pruinescence and most do not show yellow antenodal cross-veins. Skimmers have 10–14 antenodal cross-veins, the one nearest to the node being complete, whereas most of the 'small' dragonflies have fewer than ten and the one nearest the node is incomplete, crossing only as far as the sub-costa (denoted as ·5). The difference in antenodal venation between skimmers, darters and dropwings is illustrated here.

All whitefaces (*pp. 284–293*) have white a 'face', black wing bases and white veins outside the wing-spots; usually 7–8 antenodal cross-veins. Localized, mostly northern.

Comparison of antenodal cross-veins

Darters

11 species

GENUS: *Sympetrum (pp. 300–321)*

KEY FEATURES: ◆ wings on some with red venation or yellow bases ◆ usually 6·5 or 7·5 antenodal cross-veins ◆ male abdomen reddish or black ◆ female/immatures buffish with sparse dark markings ◆ perch from ground up to tip of branches ◆ breed mostly in standing waters

Identify species by abdomen and thorax colour and patterning | eye colour | leg colour | wing/vein coloration | 'face' pattern

Darter

Scarlet
1 species

GENUS: *Crocothemis* (p. 323)

KEY FEATURES: ◆ eyes blue below ◆ abdomen broad
◆ legs pale, unmarked ◆ wings with broad yellow bases and
8·5–11·5 antenodal cross-veins ◆ male all-red ◆ female/
immatures yellow-brown with pale stripes on shoulders and
between wing bases ◆ perches low in vegetation or (males)
patrol over open water ◆ breeds mostly in standing waters

Scarlet

Dropwings
4 species

GENUS: *Trithemis* (pp. 324–331)

KEY FEATURES: ◆ frons shining metallic ◆ eyes blue-grey below
◆ wings with amber base ◆ 9·5–12·5 antenodal cross-veins
◆ male red, violet-pink or blackish ◆ female yellow-brown with
dark markings ◆ often perch in obelisk position
◆ breed in standing or flowing waters

Identify species by abdomen colour and patterning | thorax
patterning | wing coloration

Dropwing

Pennant
1 species

GENUS: *Selysiothemis* (p. 332)

KEY FEATURES: ◆ small, with large head ◆ frons not shiny
◆ eyes brown above, blue below ◆ sparse, pale venation
◆ 6 antenodal cross-veins ◆ wing-spots pale
◆ male black, slightly pruinose ◆ female/immatures buffish with
dark markings ◆ perches low on tip of plant stems with broad
hindwings blowing in breeze (hence 'pennant')
◆ breeds in standing waters

Pennant

Percher
1 species

GENUS: *Diplacodes* (p. 334)

KEY FEATURES: ◆ small ◆ eyes brown above, blue below
◆ appendages white ◆ wing bases have small yellow patch
◆ 6·5 or 7·5 antenodal cross-veins ◆ wing-spots large, dark
◆ male black with frons shiny purple-black ◆ female/immatures
buffish with dark markings ◆ female/immatures pale yellowish
with black markings ◆ perches low in vegetation
◆ breeds in standing waters

Percher

Groundling
1 species

GENUS: *Brachythemis* (p. 336)

KEY FEATURES: ◆ small ◆ 6·5–8·5 antenodal cross-veins
◆ wing-spots black-and-white ◆ male blackish with broad
dark band across outerwing ◆ female/immatures pale yellowish
with dark markings ◆ typically perches on bare ground
◆ follows large mammals and vehicles ◆ breeds in standing
waters

Groundling

Identification of male darters, dropwings, pennant, percher & groundling

Key identification features of males

Identification of the red, orange, violet-pink or blackish mature males is generally fairly straightforward based on a few key features. Males have long appendages held close together, whereas those of females are short and well separated at the base. A male of each species is shown here, annotated to indicate the key identification features. Females and immature males tend to be yellowish-brown and are more challenging to identify (see the following two pages).

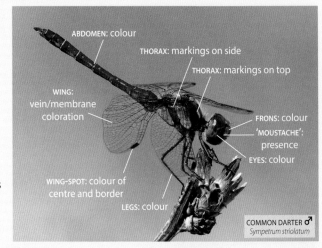

ABDOMEN: colour

THORAX: markings on side

THORAX: markings on top

WING: vein/membrane coloration

FRONS: colour

'MOUSTACHE': presence

EYES: colour

WING-SPOT: colour of centre and border

LEGS: colour

COMMON DARTER ♂
Sympetrum striolatum

Male dropwings, pennant, percher & groundling compared

A suite of species of southerly distribution, most of which have spread from Africa or Asia in recent decades in response to climate change; further range expansions are to be expected in future.

S

WING-SPOTS: small, whitish

VENATION: sparse, pale

BODY: black

HINDWING BASE: broad

Black Pennant
Selysiothemis nigra (p. 332)

SW

LEGS: pink

ABDOMEN: scarlet

HINDWING BASE: large amber patch

Orange-winged Dropwing
Trithemis kirbyi (p. 324)

S

ABDOMEN: slender, red

THORAX: top purple

S8–9: black bands

Red-veined Dropwing
Trithemis arteriosa (p. 326)

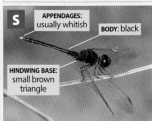

S

APPENDAGES: usually whitish

BODY: black

HINDWING BASE: small brown triangle

Black Percher
Diplacodes lefebvrii (p. 334)

S

THORAX: pinkish-violet

ABDOMEN: broad, pinkish-violet

Violet Dropwing
Trithemis annulata (p. 328)

SE

BODY: pruinose dark blue

HINDWING BASE: small brown triangle

Indigo Dropwing
Trithemis festiva (p. 330)

S

BODY: black

WINGS: dark band

Northern Banded Groundling
Brachythemis impartita (p. 336)

Male darters & scarlet compared

● = widespread; **C** = central; **E** = eastern;
S = southern/Mediterranean; **W** = western.

'MOUSTACHE': absent

THORAX: top brown; side with 2 yellow panels

Common Darter
Sympetrum striolatum (p. 302)

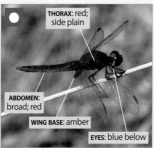

THORAX: red; side plain

ABDOMEN: broad; red

WING BASE: amber

EYES: blue below

Broad Scarlet
Crocothemis erythraea (p. 322)

THORAX: black

ABDOMEN: waisted; black

Black Darter
Sympetrum danae (p. 300)

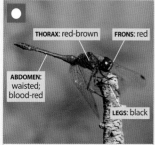

THORAX: red-brown   FRONS: red

ABDOMEN: waisted; blood-red

LEGS: black

Ruddy Darter
Sympetrum sanguineum (p. 312)

C/E

'MOUSTACHE': present

THORAX: side poorly marked

Moustached Darter
Sympetrum vulgatum (p. 304)

C/E

ABDOMEN: black 'wedge'-shapes on side   LEGS: black

Spotted Darter
Sympetrum depressiusculum (p. 314)

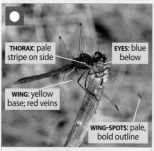

THORAX: pale stripe on side   EYES: blue below

WING: yellow base; red veins

WING-SPOTS: pale, bold outline

Red-veined Darter
Sympetrum fonscolombii (p. 316)

C/E

ABDOMEN: orange-red; black lower edge

WING BASE: large amber patch

Yellow-winged Darter
Sympetrum flaveolum (p. 318)

C/E

ABDOMEN: red

WINGS: dark band

Banded Darter
Sympetrum pedemontanum (p. 320)

C/S

S8–9: usually unmarked

THORAX: side poorly marked

LEGS: yellowish-brown

Southern Darter
Sympetrum meridionale (p. 308)

SW

SIZE: large

THORAX: side with 2 yellow panels

Island Darter
Sympetrum nigrifemur (p. 306)

SW

S2–3: black line on side

EYES: blue-grey below

Desert Darter
Sympetrum sinaiticum (p. 310)

Identification of female darters, dropwings, pennant, percher & groundling

Key identification features of females
Female and immature male dragonflies in this group are generally yellowish-brown and are typically rather nondescript, although some have conspicuous black markings. They can be difficult to identify as it is often hard to see certain key features, such as the colour of the legs and wing-spots, and the extent of black on the thorax and down the sides of the 'face'. A female of each species is shown here, annotated to indicate the key identification features. Some females may take on the colour of males, particularly with age (males are covered on the previous two pages).

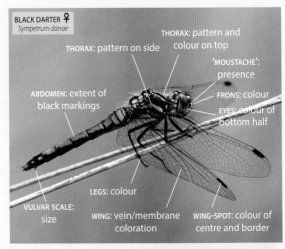

BLACK DARTER ♀
Sympetrum danae

THORAX: pattern and colour on top

THORAX: pattern on side

'MOUSTACHE': presence

FRONS: colour

ABDOMEN: extent of black markings

EYES: colour of bottom half

LEGS: colour

VULVAR SCALE: size

WING: vein/membrane coloration

WING-SPOT: colour of centre and border

Female dropwings, pennant, percher & groundling compared

A suite of species of southerly distribution, most of which have spread from Africa or Asia in recent decades in response to climate change; further range expansions are to be expected in future.

S

ABDOMEN: black line down centre

WING-SPOTS: small, whitish

VENATION: sparse, pale

Black Pennant
Selysiothemis nigra (p. 332)

All ♀ dropwings
EYES: blue-grey below

SW

TIBIAE: yellow

HINDWING BASE: amber patch

Orange-winged Dropwing
Trithemis kirbyi (p. 324)

S

THORAX: 2 black lines across side

S8–9: black bands

ABDOMEN: continuous black along side

Red-veined Dropwing
Trithemis arteriosa (p. 326)

S

ABDOMEN: thick black line down centre

APPENDAGES: pale

HINDWING BASE: small brown triangle

Black Percher
Diplacodes lefebvrii (p. 334)

S

THORAX: dark line across side, extending to S3/4

HINDWING BASE: amber patch

Violet Dropwing
Trithemis annulata (p. 328)

SE

WINGTIP: amber

ABDOMEN: black lines down centre and side

Indigo Dropwing
Trithemis festiva (p. 330)

S

ABDOMEN: black lines down centre and side

WING-SPOTS: cream, outer third dark

EYES: 3 brown stripes across top

Northern Banded Groundling
Brachythemis impartita (p. 336)

Female darters & scarlet compared

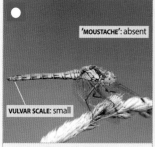

'MOUSTACHE': absent

VULVAR SCALE: small

Common Darter
Sympetrum striolatum (p. 302)

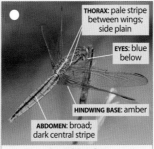

THORAX: pale stripe between wings; side plain

EYES: blue below

HINDWING BASE: amber

ABDOMEN: broad; dark central stripe

Broad Scarlet
Crocothemis erythraea (p. 322)

THORAX: black triangle on top; 3 yellow spots in black panel on side

LEGS: black

ABDOMEN: continuous black along side

Black Darter
Sympetrum danae (p. 300)

THORAX: top with black 'T'

'MOUSTACHE': present

LEGS: black

Ruddy Darter
Sympetrum sanguineum (p. 312)

C/E

'MOUSTACHE': present

VULVAR SCALE: prominent

Moustached Darter
Sympetrum vulgatum (p. 304)

C/E

LEGS: black

ABDOMEN: black 'wedge'-shapes on side

Spotted Darter
Sympetrum depressiusculum (p. 314)

WING-SPOTS: pale, bold outline

INNERWING: yellow veins

THORAX: side dull greenish-yellow

EYES: blue below

Red-veined Darter
Sympetrum fonscolombii (p. 316)

C/E

ABDOMEN: 2 black lines along side

HINDWING BASE: extensive amber

Yellow-winged Darter
Sympetrum flaveolum (p. 318)

C/E

WING-SPOTS: whitish

WING: dark band

Banded Darter
Sympetrum pedemontanum (p. 320)

C/S

THORAX: side poorly marked

LEGS: yellowish-brown

Southern Darter
Sympetrum meridionale (p. 308)

SW

'MOUSTACHE': absent

SIZE: large

Island Darter
Sympetrum nigrifemur (p. 306)

SW

THORAX: side poorly marked

EYES: blue-grey below

S2–3: black line on side

Desert Darter
Sympetrum sinaiticum (p. 310)

LC Black Darter

Sympetrum danae × 1·5

Common	=
Bog, pond, lake	

J F M A M J J A S O N D

Overall length:	29–34 mm
Hindwing:	20–30 mm

One of the smallest darters and the only *Sympetrum* with no red on the male. A very active late summer species of boggy uplands and lowland heathland.

Identification: Small. Both sexes have **black** legs and wing-spots and very **broad** base to hindwing; thorax has yellow side separated by thick black panel in which are **three yellow spots** (also a feature of dark Common Darters (*p. 302*) in NW). ♂ Mainly **black** on top of eyes, thorax and abdomen (superficially similar small black species occur only in S); abdomen has small yellow marks on the side and **distinct 'waist'** around S4; wings clear. ♀ Blacker than other darters; thorax brown on top enclosing **black triangle**; abdomen broad-based, tapering, yellow-ochre with **black inverted 'T'-shapes** on S8–9 and **continuous broad black** markings along side; vulvar scale **prominent**, projecting out at right angle (as Moustached Darter (*p. 304*)); wing bases have small yellow patches; eyes brown above. VARIATION: Immatures as female, but yellower; with age, yellow areas darken to brown, resulting in males appearing black.

Behaviour: Often basks on bare ground and pale objects. Flight action rather skittish. Males not strongly territorial but actively seek out females. May disperse from breeding sites and turn up with other migrating darters.

Breeding habitat: Shallow, fish-free acidic pools, lake margins and ditches, usually with abundant bog-mosses (*Sphagnum* spp.) and fringing rushes (*Juncus* spp.) and sedges (*Carex* spp.); rarely at alkaline sites. Scarce in extreme N and S of range; only in mountains in S.

♂ 'face' yellowish, becoming darker with age (darker than whitefaces (*pp. 284–293*))

LOOK-ALIKES (in range)
Other darters (*pp. 302–321*)
Whitefaces (*pp. 284–293*)

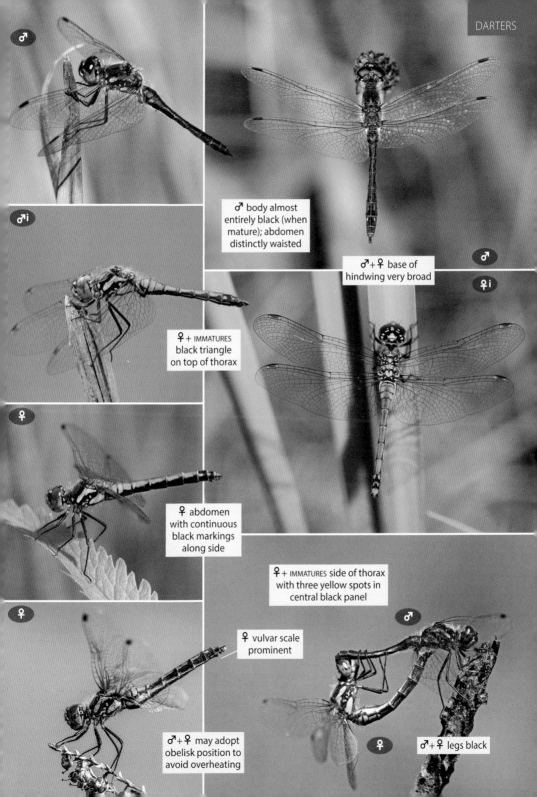

♂ body almost entirely black (when mature); abdomen distinctly waisted

♂+♀ base of hindwing very broad

♀+ IMMATURES black triangle on top of thorax

♀ abdomen with continuous black markings along side

♀+ IMMATURES side of thorax with three yellow spots in central black panel

♀ vulvar scale prominent

♂+♀ may adopt obelisk position to avoid overheating

♂+♀ legs black

LC Common Darter

Sympetrum striolatum ×1·5

Abundant =

Various, mainly standing waters

J F M A M J J A S O N D

Overall length:	35–44 mm
Hindwing:	24–30 mm

One of the most common and widespread dragonflies in Europe, which continues to fly late in the year, even into winter. Dark individuals in NW were formerly treated as a separate species, Highland Darter *Sympetrum nigrescens*, but this is no longer considered valid on the basis of DNA analysis.

Identification: Small, although larger than most darters. Both sexes have a thin black line across top of frons **not extending down sides** (unlike similar Moustached Darter (*p. 304*)); top of eyes brown; thorax brown, some with faint pale shoulder stripes; abdomen has black mark along top of S8 and S9, like most other darters; legs blackish, usually with pale **yellowish** stripe; wings clear with tiny area of yellow at base; wing-spots yellow to reddish-brown. ♂ Side of thorax has two **large yellow patches** divided by a reddish-brown panel; abdomen **orange-red** and slightly waisted. ♀ Thorax side yellow with central rectangle of thin black lines; abdomen yellow-ochre with discontinuous small black lines along side; vulvar scale protrudes less than in Moustached Darter. **VARIATION:** Immatures like bright female; yellow on female becomes dull with age; some females turn reddish, like male. Along extreme NW coasts, some have **more extensive black** down side of eyes and abdomen and on legs; three small, irregular reddish or yellowish spots in dark central panel on side of thorax (as Black Darter (*p. 300*)); and often black 'T'-shape on top of thorax in female.

Behaviour: Basks on sheltered, bare, often pale ground. Territorial males perch on or near the ground, or hover several metres above breeding waters. Feeding perches may be high in trees and shrubs. Leaves wetlands after emergence, returning to breed in late summer. Large migrations, including pairs in tandem, regular during warm airflows.

Breeding habitat: An early colonist, occupying a wide range of standing waters, even bare, shallow, newly formed and residual pools; less often in slow-flowing and brackish waters. Absent from many upland areas and the NE.

♂ + ♀ black across top of frons does not extend down side (a feature also shown by Southern (*p. 308*), Island (*p. 306*) and Desert Darters (*p. 310*))

LOOK-ALIKES (in range)
Other darters (*pp. 300–321*)
Dropwings (*pp. 324–331*)
♀ skimmers (*pp. 262–279*)

♂ thorax side with two yellow patches divided by reddish-brown panel

♂ top of thorax brown

♂ abdomen orange-red, slightly waisted

♂+♀ dark 'form' side of thorax with three yellow spots in central black panel (like Black Darter (*p. 300*), but lack black triangle on top of thorax)

♀ dark 'form'

♂+♀ wings clear with tiny area of yellow at base; wing-spots yellow to reddish-brown

♀ abdomen with discontinuous black markings along side

♀ thorax side yellow, with central rectangle outlined in black

♂+♀ legs dark, usually with yellowish stripe

♀ vulvar scale inconspicuous

LC **Moustached Darter**

Sympetrum vulgatum ×1·5

Vagrant Darter

Abundant =

Various standing waters

J F M A M J J A S O N D

Overall length:	35–40 mm
Hindwing:	24–29 mm

♂ + ♀ black across top of
frons extends down side –
the so-called 'moustache'

May be overlooked due to its similarity to Common Darter (*p. 302*); photographs or binoculars help to confirm presence of a 'moustache' and, on the female, very prominent vulvar scale.

Identification: Small. In both sexes, thin black line across top of frons **extends down sides** ('moustache', absent in very similar Common Darter (see *p. 302*)); top of eyes brown; thorax brown, some with faint pale shoulder stripes; abdomen has black mark along top of S8 and S9, like most other darters; legs blackish, usually with **yellowish** stripe; wings clear with tiny area of yellow at base; wing-spots yellow to reddish-brown. ♂ Thorax **poorly marked**, tinged reddish and lacking the two yellow patches on side shown by Common Darter; abdomen **red**, slightly waisted (more than in Common Darter) and with tiny black spots on S3–7, each surrounded by **yellow ring**. ♀ Thorax side yellow with central rectangle of thin black lines; abdomen yellow-ochre with discontinuous small black lines along side; vulvar scale prominent and protruding at a right angle. VARIATION: Immatures like bright female; yellow on female becomes dull with age; some females turn reddish, like male. Small, pale subspecies *ibericum* [not illustrated] in Spain has pale legs and costa and lacks 'moustache', rather like Southern Darter.

Behaviour: Similar to Common Darter, favouring low and often pale perch from which to investigate potential prey or mates. An occasional migrant but less prone to large-scale dispersion than other darters.

Breeding habitat: A wide range of still waters, often with more luxuriant vegetation than those favoured by Common Darter. Common in C and E Europe, more abundant than Common Darter in NE of range; rarer and mainly in upland areas in S and W. Most numerous in August.

LOOK-ALIKES (in range)
Other darters (*pp. 300–321*)
♀ skimmers (*pp. 262–279*)

♂

♂ thorax brown

♂ thorax side rather uniform, lacking yellow patches

♂

♂ abdomen red, slightly waisted

♂+♀ wings clear with tiny area of yellow at base

♀

♂ S3–7 tiny black spots, each surrounded by yellow ring

♀ abdomen with discontinuous black markings along side

♀ •

♂+♀ legs black, usually with yellowish stripe

♀ red 'form'

♀ vulvar scale prominent

E LC Island Darter

Sympetrum nigrifemur ×1·5

Locally common =

Flowing and standing waters

J F M A M J J A S O N D

Overall length:	43–48 mm
Hindwing:	29–33 mm

♂ + ♀ black across top of frons does not extend down side

Essentially a large, slightly darker version of Common Darter (*p. 302*) with minor differences in male secondary genitalia and shape of female vulvar scale. Replaces that species in Macaronesia, where it is endemic; genetic studies may show that it is a subspecies of Common Darter.

Identification: Medium-sized, the **largest darter**. As Common Darter, both sexes have a thin black line across top of frons not extending down sides (see *below* and *p. 302*); eyes and top of thorax brown, the latter with faint pale shoulder stripes; abdomen has black mark along top of S8 and S9, like most other darters; legs blackish with thin yellowish or reddish stripe; wings clear with tiny area of yellow at base; wing-spots yellow to reddish-brown. ♂ Side of thorax has two **large yellow patches** divided by a reddish-brown panel; abdomen **orange-red** and slightly waisted. ♀ Thorax side yellow marked with black lines; abdomen yellow-ochre with discontinuous small black lines along side; vulvar scale protrudes less than in Moustached Darter (*p. 304*), but more robust, blunt and less prominent than in Common Darter. VARIATION: May have more extensive black markings, like some dark Common Darters in NW Europe. Immatures like bright female; yellow on female becomes dull with age; some females turn reddish, like male.

Behaviour: Much as Common Darter but flies in all months and probably has two generations per year, with rapid larval growth in summer.

Breeding habitat: Most often seen at flowing waters, but also at various standing waters. Known from about 150 sites, with breeding confirmed on Madeira, the Selvagens and the Canary Islands (Gran Canaria, La Gomera, La Palma and Tenerife, with vagrants recorded on other islands).

LOOK-ALIKES (in range)
Red-veined Darter (*p. 316*)
Red-veined Dropwing (*p. 326*)

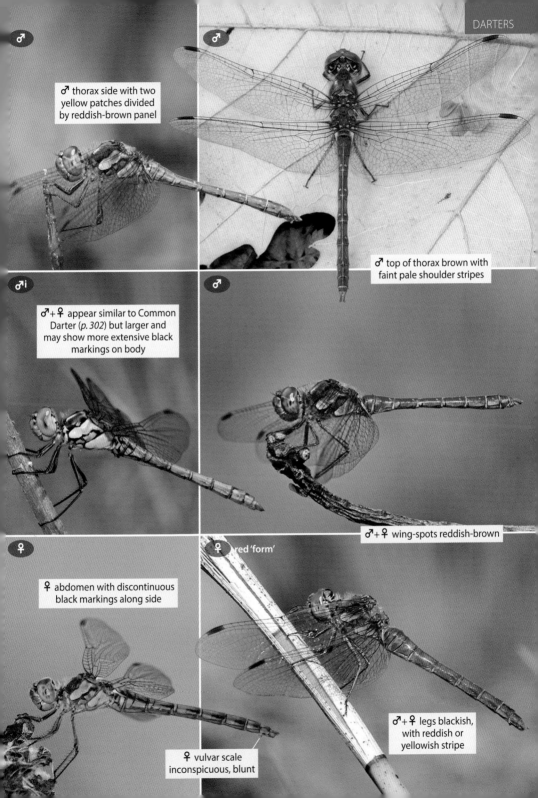

♂ thorax side with two yellow patches divided by reddish-brown panel

♂ top of thorax brown with faint pale shoulder stripes

♂i ♂ + ♀ appear similar to Common Darter (*p. 302*) but larger and may show more extensive black markings on body

♂ + ♀ wing-spots reddish-brown

♀ abdomen with discontinuous black markings along side

♀ red 'form'

♂ + ♀ legs blackish, with reddish or yellowish stripe

♀ vulvar scale inconspicuous, blunt

LC Southern Darter

Sympetrum meridionale ×1·5

Common △

Lake, pond

J F M A M J J A S O N D

Overall length:	35–40 mm
Hindwing:	25–30 mm

♂ + ♀ black across top of frons does not extend down side (as on Common Darter (p. 302))

Similar to Common Darter (*p. 302*) but with plain side to thorax and pale legs. Has spread significantly to the N in recent decades and further range expansion is likely with climate change.

Identification: Small. Very similar to Common Darter, but very little black visible at top of frons and, like that species, usually no extension down sides (see *below* and *p. 302*); thorax brown on top with faint pale shoulder stripes, side **virtually unmarked** with **two black spots**, uppermost very small; legs **yellowish-brown** with only a thin black stripe down length; wing-spots slightly larger than in Common Darter. ♂ 'Face' pink; side of thorax and legs may be tinged with red; abdomen orange-red, virtually unmarked. ♀ Thorax **pale yellow on** side; abdomen yellowish-brown with discontinuous faint black lines along side; vulvar scale small and rounded from side, barely visible. VARIATION: Immatures like bright female; yellow on female becomes dull with age; some females turn reddish, like male; wings often have a little yellow at base; black on top of frons extends down sides on some.

Behaviour: In S, often found in hills after emergence, returning to water later after autumn rains. Perches on the ground or in tall vegetation. Particularly prone to infestations of red mites on wings.

Breeding habitat: Breeds in lowlands in shallow ponds and sheltered parts of lakes with abundant emergent vegetation; can tolerate brackish conditions. Sites often dry out in summer or at least are affected by falling water levels.

LOOK-ALIKES (in range)
Other darters (*pp. 300–321*)
♀ skimmers (*pp. 262–279*)
Broad Scarlet (*p. 322*)

♂ ♂+♀ side of thorax almost unmarked, but with two small black spots

♂ usually lacks small black marks on S8 and S9 (often found on other darters)

♂+♀ wing-spots reddish-brown (slightly larger than in Common Darter (*p. 302*))

♀ i

♂+♀ legs yellowish-brown with thin black stripe

♂ i

♀ o

♀ abdomen with discontinuous black markings along side

♀ vulvar scale small, rounded

LC Desert Darter

Sympetrum sinaiticum × 1·5

Uncommon, locally
common =

Various standing and
flowing waters

J F M A M J J A S O N D

Overall length:	34–37 mm
Hindwing:	24–29 mm

Intermediate in appearance between Common Darter (*p. 302*) and Southern Darter (*p. 308*), but with grey underside to eyes and black bars on the side of S2–3. Its localized distribution comprises parts of S and E Spain, N Africa and the Levant.

Identification: Small. Very similar to Common and Southern Darters, both sexes having **blue-grey** underside to eyes (instead of green or brown), with no black down sides (see *p. 302*); thorax brown on top with faint pale shoulder stripes and side **poorly marked** with very thin black lines; abdomen with diagnostic **black line** on sides of S2–3; legs yellowish with dark stripe down length. ♂ 'Face' pink; thorax brownish; abdomen not waisted, pinkish-red, rather plain, often yellow along side and, at most, has only faint black mark along top of S8 and S9. ♀ Thorax **plain yellow** on side; abdomen yellowish-brown with black lines along sides of S4–8; vulvar scale not obvious, between those of Common and Southern Darter in size. VARIATION: Immatures like bright female; yellow on female becomes dull with age; some females turn reddish, like male; extent of black may vary; wings may have a little yellow at base.

Behaviour: Gathers in trees and bushes away from water after emergence, returning to breed in October–December after autumn rains.

Breeding habitat: Permanent and perhaps mainly temporary ponds, lakes, streams and ditches in arid areas; irrigation tanks and other artificial waters often used.

♂ + ♀ black across top of
frons extends down sides

LOOK-ALIKES (in range)
Other darters (*pp. 300–321*)
♀ skimmers (*pp. 262–279*)

310

♂

♂

♂+♀ underside
of eyes blue-grey

♀

♂

♂+♀ S2–3 with
diagnostic black
line on side

♂+♀ legs
yellowish with
dark stripe

♀

♀ red 'form'

♂+♀ side of thorax rather
plain, with very thin black lines

♀ vulvar scale
inconspicuous

LC Ruddy Darter

Sympetrum sanguineum ×1·5

Abundant =

Lake, pond, canal, ditch

J F M A M J J A S O N D

Overall length:	34–39 mm
Hindwing:	23–31 mm

♂ + ♀ black across top of
frons extends down sides

The commonest black-legged darter, the deep-red males also having a bright red 'face' and obviously waisted abdomen. Common across much of its range, becoming scarce in, or absent from, far N and S.

Identification: Small. Both sexes have a thin black line across top of frons extending **down sides** (see *below*); top of thorax marked with dark 'T'-shape; abdomen shorter than Common Darter, with thick black marks along lower edge; small area of yellow at wing bases; wing-spots red-brown; legs **black**. ♂ Eyes **red-brown** on top, dark green below; frons red; thorax rather plain **reddish-brown**; abdomen **waisted** and **blood-red** with 'clubbed' tip and thick black line along top of S8 and S9; appendages reddish. ♀ Thorax **yellow-ochre** with single thin black line running from hind leg to hindwing base; abdomen yellow-ochre; eyes brown above, yellow below; frons yellow; vulvar scale small, insignificant. VARIATION: Immatures like bright female; yellow on female becomes dull (turning reddish in some) with age; wings may have more extensive yellow suffusion, typically so in both sexes in SW Turkey (wing bases and around nodes).

Behaviour: Often found in tall vegetation slightly back from the water's edge. Flight jerky, hovering more than Common Darter. Habitually lays eggs in tandem among dense emergent plants and onto wet mud. Sometimes disperses with other darters.

Breeding habitat: Well-vegetated ponds, lakes, canals and ditches, often near woodland; sometimes sluggish rivers and brackish waters but rarely in acidic waters; can breed in swampy areas that dry out in late summer.

LOOK-ALIKES (in range)
Other darters (*pp. 300–321*)
Broad Scarlet (*p. 322*)
♀ skimmers (*pp. 262–279*)

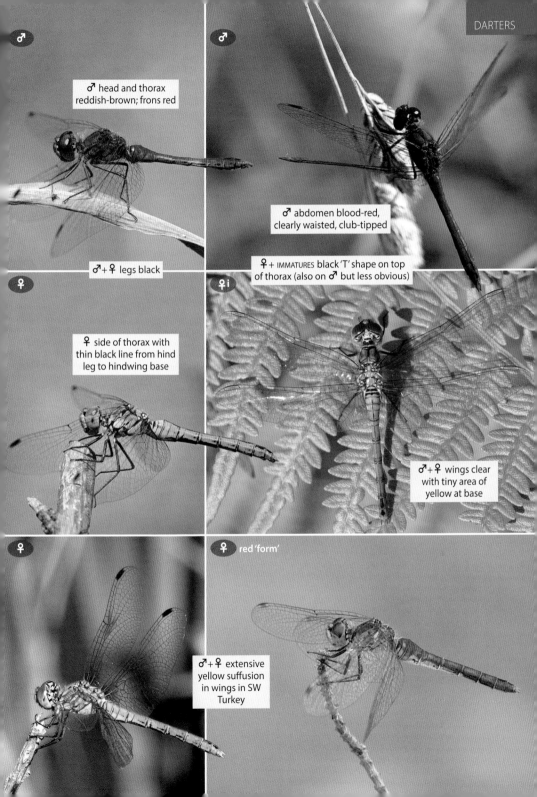

♂ ♂

♂ head and thorax reddish-brown; frons red

♂ abdomen blood-red, clearly waisted, club-tipped

♂ + ♀ legs black

♀ + IMMATURES black 'T' shape on top of thorax (also on ♂ but less obvious)

♀

♀ side of thorax with thin black line from hind leg to hindwing base

♀i

♂ + ♀ wings clear with tiny area of yellow at base

♀

♀ red 'form'

♂ + ♀ extensive yellow suffusion in wings in SW Turkey

VU Spotted Darter

Sympetrum depressiusculum ×1·5

Marshland Darter

Scarce and local ▽

Seasonal swamps

J F M A M J J A S O N D

Overall length:	29–34mm
Hindwing:	24–28mm

♂+♀ **black across top of frons extends down sides**

One of the smallest and scarcest darters. A significant decline in the late 20th century continues and it is now found patchily in lowland temperate areas, although sometimes locally abundant.

Identification: Small. Both sexes have eyes that are brown above, green below; frons yellow with **thick** black line across top extending **down sides** (see *below*); thorax with black 'collar' across front of thorax; abdomen with **diagnostic black 'wedge'-shapes** on sides of S3–8, also visible from above, and black line along top of S8 and S9; outerwing with **many small cells** at rear edge (see *below*); wing-spots relatively large and orange, **boldly outlined** in black; legs **black**. ♂ Thorax brown on top, yellow-brown on side; abdomen flattened and slightly swollen near tip, **orange-red**, often with yellow side; appendages yellowish. ♀ Thorax brown on top, yellow with black lines on side; abdomen yellow-ochre; vulvar scale small. **VARIATION:** Immatures are like bright female.

Behaviour: Flight weak and fluttery; a poor disperser. May form large communal roosts.

Breeding habitat: Swamps and other lowland shallow wetlands that are typically dry in winter and flooded in spring (by snow melt, for example), followed by lush growth of tall emergent plants. Formerly common in temporarily drained areas such as carp breeding ponds and rice fields.

♂+♀ venation along rear edge of outerwing much denser in Spotted Darter than in Ruddy Darter

SPOTTED DARTER

RUDDY DARTER

LOOK-ALIKES (in range)
Other darters (*pp. 300–321*)
♀ skimmers (*pp. 262–279*)

♂ face yellow

♂ abdomen orange-red, often with yellow side

♂ appendages yellow

♂ + ♀ wing-spots orange, boldly outlined in black

♂ + ♀ side of abdomen with diagnostic black 'wedge'-shapes on side (also visible from above)

♂ + ♀ legs black

♀ vulvar scale small

LC Red-veined Darter

Sympetrum fonscolombii ×1·5

Common △
Various standing waters

J F M A M J J A S O N D

Overall length:	33–40 mm
Hindwing:	26–31 mm

♂ + ♀ black across top of
frons extends down sides

A colourful, highly dispersive darter that has spread N in recent decades, aided by climate change. Most abundant around the Mediterranean.

Identification: Small, but relatively large for a darter. Both sexes have eyes that are **blue** below; black line across top of frons extends **down sides** (see *below*); thorax with black 'collar' across front of thorax; abdomen virtually parallel-sided with thin black line on side and thick black line along top of S8 and S9; wings may appear blue-grey at times, hindwings with obvious **yellow base**, less extensive than in Yellow-winged Darter (*p. 318*); wing-spots **pale, boldly outlined in black**; legs black with yellow stripe. ♂ Eyes red-brown above; frons and thorax **reddish**, the latter with **pale stripe** across side; abdomen mostly **pinkish-scarlet** above; costa and other veins in inner half of wings conspicuously **red** (brighter than in other darters). ♀ Eyes brown above; thorax and abdomen yellow-ochre, the former with thin black lines on **dull greenish-yellow** side; wings have extensive **yellow veins** in basal half; vulvar scale small, insignificant. VARIATION: Immatures like bright female; yellow on female becomes dull with age; some females turn reddish, like male; extent of yellow suffusion on wings variable. Legs may be all-black (*e.g.* in Azores, where individuals are darker).

Behaviour: Males highly territorial, spending more time in flight than perched. Flight strong and erratic, with frequent hovering, often far out over water. Two generations per year where temperatures are higher (typical in S), the first producing red males earlier in summer than any other darter. Disperses after emergence, resulting in periodic large migrations and lack of persistence at some breeding sites.

Breeding habitat: Warm, lowland standing waters, often bare and shallow, including brackish wetlands, dune ponds, quarries, rice fields and newly created wetlands; occasional in slow-flowing rivers. Scarcer in N of range, where populations supplemented by dispersal from S.

LOOK-ALIKES (in range)
Other darters (*pp. 300–321*)
Dropwings (*pp. 324–331*)
♀ skimmers (*pp. 262–279*)
Broad Scarlet (*p. 322*)

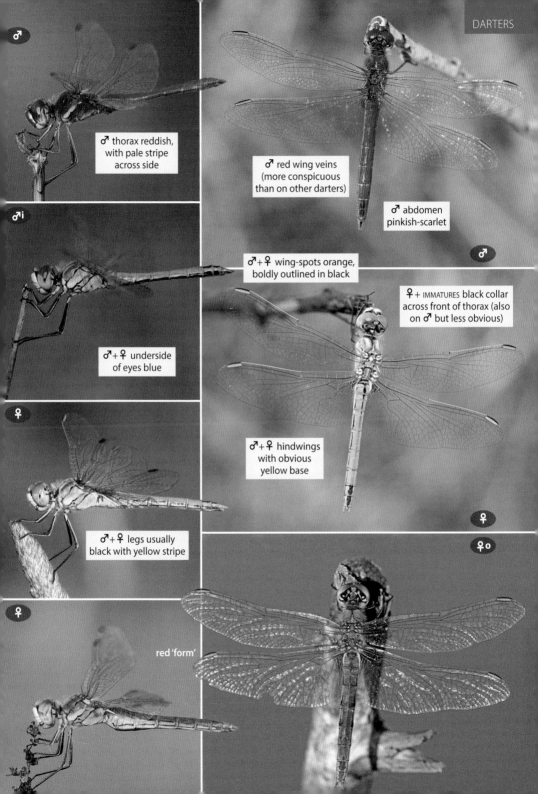

♂ thorax reddish, with pale stripe across side

♂i

♂ red wing veins (more conspicuous than on other darters)

♂ abdomen pinkish-scarlet

♂

♂+♀ wing-spots orange, boldly outlined in black

♂+♀ underside of eyes blue

♀+ IMMATURES black collar across front of thorax (also on ♂ but less obvious)

♀

♂+♀ hindwings with obvious yellow base

♂+♀ legs usually black with yellow stripe

♀

♀o

red 'form'

LC Yellow-winged Darter

Sympetrum flaveolum ×1·5

Common =

Lake, pond, bog

J F M A M J J A S O N D

Overall length:	32–37 mm
Hindwing:	23–33 mm

♂ + ♀ black across top of frons extends down sides

A rather small, colourful darter with a somewhat erratic flight.

Identification: Small, both sexes smaller than Common Darter (*p. 302*). Eyes yellowish below; black line across top of frons extends **down sides** (see *below*); thorax with black 'collar' across front, otherwise relatively unmarked; abdomen virtually parallel-sided with **continuous black** marking along side and black line along top of S8 and S9; wings have **orange** suffusion and **bright yellow** veins over much of the basal half, more extensive than in Red-veined Darter (*p. 316*); wing-spots **red-brown** with **thick black margins**; legs black with yellow stripe. ♂ Frons red; thorax **reddish-brown**; abdomen **orange-red** with **thick black lower edge**; wings with **yellow** costa. ♀ Thorax and abdomen yellow-brown, the former with thin black lines on side, the latter with **two continuous** black lines along side; wings with **yellow** extending **beyond node**; vulvar scale small, insignificant. VARIATION: Immatures like bright female; yellow on female becomes dull and yellow suffusion on wings fades with age.

Behaviour: Flight rather weak, fluttering and erratic, with frequent hovering. Perches frequently, often part-way up tall vegetation. Territorial males fairly aggressive. May be seen egg-laying over dried-out grassy areas.

Breeding habitat: Well-vegetated standing waters in cool temperate regions, including shallow lake margins, flood meadows and ponds in dunes, fens and peat bogs, which at least partly dry out in summer. Mostly short-lived populations in mountains of S and W, especially after irregular migrations.

LOOK-ALIKES (in range)
Other darters (*pp. 300–321*)
♀ skimmers (*pp. 262–279*)

♂

♂

♂ thorax reddish-brown

♂ costa yellow

♂ abdomen orange-red with continuous thick black lower edge

♂+♀ wings with orange suffusion and yellow veins over basal half

♂+♀ wing-spots red-brown, outlined in black

♀

♀

♀ two black lines along side of abdomen

♂+♀ legs black with yellow stripe

♂i

♀i

LC **Banded Darter**

Sympetrum pedemontanum ×1·5

Locally common **=**

Slow-flowing and standing waters

J F M A M J J A S O N D

Overall length:	28–35 mm
Hindwing:	21–28 mm

♂ + ♀ thin black line across top of frons extends down sides

This attractive species resembles a Ruddy Darter (*p. 312*), but with conspicuous dark wing bands. In Europe, only the more southerly and sombrely coloured Northern Banded Groundling (*p. 336*) has similar wing bands. Mostly very localized, but has spread in recent decades.

Identification: One of the smallest darters; both sexes have a thin black line across top of frons extending down sides (see *below*), distinctive **dark brown band** near tip of each wing extending as far as (large) wing-spots, and **black legs**. ♂ Eyes red-brown above; frons red; thorax rather plain **reddish-brown**; abdomen **deep red** with slightly clubbed tip; wing-spots **pinkish-red**. ♀ Frons yellowish; thorax brown on top, yellow on side, poorly marked; abdomen yellow-ochre with black mark along top of S8 and S9; eyes brown above, yellow below; wing-spots **whitish**; vulvar scale small, insignificant. VARIATION: Immatures as female, wing bands developing soon after emergence; wing bands variable in extent; some females turn reddish, like male.

Behaviour: Weak, low, fluttering flight like Spotted Darter (*p. 314*), with which it is sometimes found. Surprisingly unobtrusive and not easily seen unless disturbed. May perch close to ground or on tips of sedge (*Carex* sp.) or rush (*Juncus* sp.) stems. Occasionally wanders far.

Breeding habitat: Slow-flowing or still waters, including pools, swamps, marshy meadows and drainage channels with little open water and emergent vegetation that is not particularly tall or dense. Breeds in floodplains inundated by snow melt, sites that dry out in either summer or winter, and seepage-fed artificial wetlands, including rice rields and irrigation channels.

LOOK-ALIKES (in range)

Other darters (*pp. 300–319*))
Northern Banded Groundling (*p. 336*)

♂ + ♀ legs black

♂ wing-spots
pinkish-red

♂ + ♀ wings
with dark brown
band near tip

♀ + IMMATURES
wing-spots
whitish

LC Broad Scarlet

Crocothemis erythraea ×1·5

Scarlet Darter

Common △	
Lake, pond	

J F M A M J J A S O N D

Overall length:	36–45 mm
Hindwing:	23–33 mm

Males of this sun-loving species are among the most vividly red of all European dragonflies. Occurs commonly in S and C Europe (as well as in Africa) and has spread N in recent decades.

Identification: Medium-sized, more robust than other darters; underside of eyes **blue**; side of thorax **plain**; abdomen **flattened, broad** in middle; wing bases have amber patches, largest and emphasized by yellow veins on hindwing; wing-spots orange-yellow **outlined strongly** in black, longer than those of darters; legs **lack** dark stripes. ♂ 'Face', top of eyes and abdomen **bright red**; thorax **reddish-brown**; veins at front of wings **red**; legs **orange-red**. ♀ Body yellow-brown; shoulder stripes **pale**; conspicuous **pale stripe** between wing bases; abdomen has narrow dark line along top of S3–10 and more-or-less distinct brown stripes down side of abdomen; vulvar scale **protrudes** conspicuously; legs **yellow-brown**. VARIATION: Immatures yellow-brown, like female; male may have dark central line on S8–10; female becomes dull olive-brown with age and may turn red like male.

Behaviour: Males are aggressively territorial at breeding sites, perching low like darters with wings held well forward. A strong flier, sometimes moving long distances and may be attracted to lights at night. 1–2 generations per year in N, potentially up to four in S.

Breeding habitat: Wide range of mainly standing, well-vegetated waters, including rice fields and temporary waters; tolerates brackish conditions and moderate levels of nutrient enrichment.

LOOK-ALIKES (in range)
Darters (*pp. 300–321*)
Orange-winged Dropwing
(*p. 324*)
Violet Dropwing (*p. 328*)
♀ skimmers (*pp. 262–279*)

♂

♂+♀ legs uniformly pale (no black)

♂ abdomen broad

♀

♂+♀ side of thorax plain, lacking dark markings

♀+ IMMATURES pale stripe between base of wings

♀ vulvar scale protrudes conspicuously

♂+♀ underside of eyes blue

♀ red 'form'

♂i

♂+♀ wing-spots orange-yellow, outlined in black

♂+♀ wings with amber patch at base, with yellow veins on hindwing

NE Orange-winged Dropwing

Trithemis kirbyi ×1·5

Kirby's Dropwing

Scarce and local △

Various standing and slow-flowing waters

J F M A M J J A S O N D

Overall length:	30–34 mm
Hindwing:	23–29 mm

Common throughout most of Africa, this is the most recent European dragonfly colonist, with the first proof of breeding being in Spain in 2008.

Identification: Small, with darter-like (*pp. 300–321*) structure; underside of eyes blue-grey (as in Broad Scarlet (*p. 322*), Red-veined Darter (*p. 316*) and Desert Darter (*p. 310*)), but smaller and slimmer and with small, dark wing-spots. ♂ body **extensively bright scarlet**; abdomen slightly waisted with black 'dash' on top of S9 (also on S8 on some); wing base has **large amber patch** covering almost one-third of wing; wing veins extensively **red**; legs **pink**. ♀ Body yellow-brown; thorax has **broad pale** shoulder stripes and irregular thin dark lines on side; abdomen has dark flecks along top, most prominent on S8–9, and more-or-less distinct discontinuous dark stripe down side, especially bold on S6–9; amber wing patches emphasized by yellow veins, less extensive than in male and **blotchy**, often with discrete rounded patch to rear of hindwing; tibiae **yellowish**. VARIATION: Immatures yellow-brown, like female; male may have dark central line on S8–10; female body and wing markings vary in intensity; body becomes dull olive-brown with age and top of thorax and abdomen may turn reddish.

Behaviour: Males often perch on waterside rocks, but are also attracted to concrete surfaces, such as those at ornamental ponds, swimming pools and drinking troughs. Females are elusive, spending much time away from water.

Breeding habitat: Shallow standing and slow-flowing waters in arid areas; typically sparsely vegetated with stony or rocky bottom, including natural and artificial ponds, lakes, rivers, streams and ditches, both permanent and temporary; can tolerate presence of fishes.

LOOK-ALIKES (in range)
Other dropwings (*pp. 326–331*)
Broad Scarlet (*p. 322*)
Darters (*pp. 300–321*)
♀ skimmers (*pp. 262–279*)

♀ red 'form'

♂

♂+♀ underside of eyes blue-grey

♂ thorax and abdomen bright scarlet

♂ legs pink

♀ tibiae yellow

♀

♂i

♂+♀ wing-spots dark

♀ amber patch at base of each wing (but less extensive than on ♂)

♀

♂

♀ broad pale shoulder stripes

♂ large amber patch at base of each wing

NE Red-veined Dropwing

Trithemis arteriosa ×1·5

Scarce and very local ?

Various standing and slow-flowing waters

J F M A M J J A S O N D

Overall length:	32–38 mm
Hindwing:	23–30 mm

Common in Africa, but scarce in Europe, being found only in the extreme SW and SE.

Identification: Small, with **slender** abdomen, similar to Red-veined Darter (*p. 316*) in some respects. In both sexes, underside of eyes **blue-grey**; thorax and abdomen **boldly marked in black**; legs **black**. ♂ Frons, top of eyes, thorax and abdomen **scarlet**; thorax **red-purple** with broad black lines across side; abdomen has **black across S8–9** and along side increasingly prominent towards tip, **continuous on S6–9**; wing bases have amber patches, smaller on forewing and much smaller than on Orange-winged Dropwing (*p. 324*); appendages black; wing veins extensively **red**; wing-spots dark. ♀ Eyes brown above; thorax yellowish with indistinct pale shoulder stripes, **pale stripe** between wing bases and pale side, the latter crossed by irregular thick black lines; abdomen yellowish, **continuous black** on side with thick 'wedge' shapes on S4–8 and split into two lines at base; S9 mostly black; amber wing bases less extensive than in male, emphasized by yellow veins; wing-spots yellowish with thick black front edge. VARIATION: Immatures yellow-brown, male with extensive orange wing venation; extent of black markings and amber in wings varies, some females having amber at nodes and tips.

Behaviour: Males perch conspicuously on rocks and emergent waterside vegetation.

Breeding habitat: Permanent and temporary standing and slow-flowing waters in arid areas; sites include ponds, lakes, gravel pits and streams.

LOOK-ALIKES (in range)
Other dropwings (*pp. 324–331*)
Darters (*pp. 300–321*)
♀ skimmers (*pp. 262–279*)

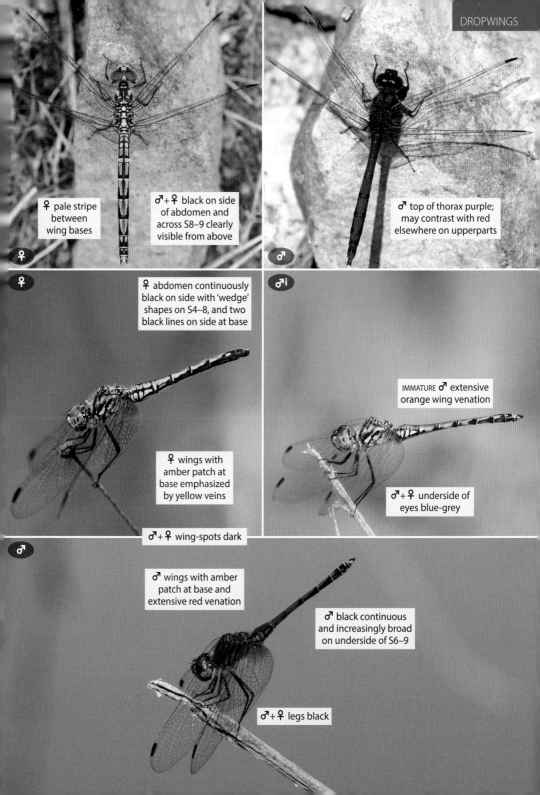

♀ pale stripe between wing bases

♂+♀ black on side of abdomen and across S8–9 clearly visible from above

♂ top of thorax purple; may contrast with red elsewhere on upperparts

♀ abdomen continuously black on side with 'wedge' shapes on S4–8, and two black lines on side at base

IMMATURE ♂ extensive orange wing venation

♀ wings with amber patch at base emphasized by yellow veins

♂+♀ underside of eyes blue-grey

♂+♀ wing-spots dark

♂ wings with amber patch at base and extensive red venation

♂ black continuous and increasingly broad on underside of S6–9

♂+♀ legs black

LC Violet Dropwing

Trithemis annulata ×1·5

Violet-marked Darter

Common △

Various standing and slow-flowing waters

J F M A M J J A S O N D

Overall length:	32–38 mm
Hindwing:	20–35 mm

Common over much of Africa and, assisted by climate warming, this spectacularly colourful species has colonized S Europe dramatically since the 1970s, spreading farther N than any other dropwing.

Identification: Small; eyes red-brown above, **blue-grey** below; abdomen rather **broad**, suggesting Broad Scarlet (*p. 322*), but smaller and with reddish wing-spots; hindwing bases have **dark amber patches**, smaller than in Broad Scarlet and Orange-winged Dropwing (*p. 324*). ♂ 'Face' **metallic purple** at top; body extensively **pruinose pinkish-violet**; abdomen has more-or-less vague darker central line and thicker black dashes on top of reddish S8–9; wing veins extensively **red**, yellow in amber patches; legs **black with violet tint**. ♀ Body yellow-brown; thorax has **broad pale** shoulder stripes, pale line between wing bases and **thick** dark lines across side; abdomen has black line along top, **splitting** into two lines on S3–7 and re-joining as a thick mark on S8–9; also black line on side of S1–3/4, continuing as thick horizontal line across side of thorax; wing veins extensively orange-red; legs black. VARIATION: Immatures like female but with orange wing veins and yellow base to legs; males turn orange and red before becoming pruinose, when central markings on abdomen become ill-defined; female becomes dull olive-brown with age, abdomen may turn reddish, and dark markings and amber in wings fade.

Behaviour: Males perch conspicuously on waterside plants, adopting the obelisk position in hot sun.

Breeding habitat: A wide range of sunny, standing and slow-flowing waters, including natural and artificial ponds, lakes, rivers, streams and ditches.

LOOK-ALIKES (in range)
Other dropwings (*pp. 324–331*)
Broad Scarlet (*p. 322*)
Darters (*pp. 300–321*)
♀ skimmers (*pp. 262–279*)

328

♀

♂+♀ underside of
eyes blue-grey

♂ distinctive pinkish-violet
pruinescence when mature

♀

♀+ IMMATURES dark
line extends forwards
from side of S3/4
across side of thorax

♂+♀ dark amber patch
at base of hindwings

♀ o

♂

♀ darkens with
age and may
become reddish

♂+♀ wing-spots reddish

♀

♂i

♀ paired dark lines
along top of S3–7

♂+♀ legs black

LC Indigo Dropwing

Trithemis festiva ×1·5

Very locally common =

Stream, river

J F M A M J J A S O N D

| **Overall length:** | 31–37 mm |
| **Hindwing:** | 23–32 mm |

This dark species is the scarcest dropwing in Europe, being found only in the extreme SE, although its range extends across S Asia, where it is common.

Identification: Small. Both sexes have brown eyes that are **blue-grey** below; abdomen slender; legs **black**; wing veins and rather short wing-spots **black**. ♂ Top of 'face' **metallic purple**; thorax and abdomen **pruinose dark blue**; appendages black; hindwing has small, dark brown triangle at base. ♀ Body roughly **equally black and yellow**; thorax has pale shoulder stripes and **thick** black lines on top and side; abdomen marked rather like female Red-veined Dropwing (*p. 326*), with **continuous, thick black lines** along top and side, the latter extending onto top at rear edge of S3–7 in increasing thickness and enclosing yellow streaks on side of S1–8, top of S8–10 all-black; hindwing bases with **small** yellow patch; wingtips usually **dusky amber**, extending to inner edge of wing-spot. VARIATION: Immatures as female; some males have pairs of orange-yellow streaks down the shiny black abdomen until fully pruinose, and/or dusky wingtips.

Behaviour: Males perch conspicuously on bankside stems or rocks.

Breeding habitat: Sunny, often fast-flowing, rocky streams and small rivers; tolerates some degree of pollution and habitat degradation.

LOOK-ALIKES (in range)
Other dropwings (*pp. 324–331*)
Black Pennant (*p. 332*)
Darters (*pp. 300–321*)
♀ skimmers (*pp. 262–279*)

♀ wingtips usually dusky amber

♀ small yellow patch at base of hindwing

♂ often has paired yellow markings on abdomen until fully mature

♀ body equally black and yellow; abdomen with continuous thick black lines along top and side

♂+♀ wing-spots black

♂ small dark brown triangle at base of hindwing

♂+♀ underside of eyes blue-grey

♂ top of 'face' metallic purple

♂ thorax and abdomen pruinose dark blue

⟨LC⟩ Black Pennant

Selysiothemis nigra ×1·5

Scarce and local △

Pond, lake

J F M A M J J A S O N D

Overall length:	30–38 mm
Hindwing:	24–27 mm

One of three small dragonflies in S Europe in which males become black (the others being Black Percher (*p. 334*) and Northern Banded Groundling (*p. 336*)); best differentiated by their wings and colour of appendages. The only species in its genus, it is very localized, having a scattered distribution extending to N Africa and S Asia.

Identification: Small, but with large head. In both sexes eyes brown above, blue-grey below; clear, **broad** wings (especially hindwing) with relatively **few veins**, mostly **very pale**; and **short, pale** wing-spots with bold black line along front and rear. ♂ Body **all-black** when fully mature, including appendages; abdomen slender. ♀ Body pale buff with bold black markings; side of thorax has three thick black diagonal lines; abdomen has irregular black markings along top, widest at S1 and S8–10, with pale bands across segment divisions; legs pale yellow at base. VARIATION: Immatures as female, although males have more extensive black markings; with age, pale areas on female darken and males may develop slight bluish pruinescence.

Behaviour: Often found a short distance from water, perched on stiff stems in open areas; wings are held slightly raised and flutter in the wind. Males patrol at waist-height over open water, hovering frequently while searching for females. Highly nomadic and may be found away from known breeding sites.

Breeding habitat: Shallow, often well-vegetated ponds and lakes, including brackish and artificial sites; sometimes in deeper, bare waters, even concrete tanks; typically in dry, often coastal, areas; strongest colonies in SE, uncommon in SW.

LOOK-ALIKES (in range)
Darters (*pp. 300–321*)
Indigo Dropwing (*p. 330*)
Black Percher (*p. 334*)
Northern Banded Groundling
(*p. 336*)

♂

♂

♂ body entirely black,
some slightly pruinose

♂+♀ wing veins
sparse and mostly pale

♀ o

♀ darkens
with age

♀

♂+♀ wing-spots
whitish with bold
black line along front
and rear, appearing
as an '=' symbol

♀

♂i

♀ side of thorax
with three diagonal
black lines

Darters, *etc.* compared *pp. 296–299*

LC Black Percher

Diplacodes lefebvrii   ×1·5

Scarce and local =

Pond, lake, marsh

J F M A M J J A S O N D

Overall length: 25–33 mm
Hindwing: 19–29 mm

A small, rather inconspicuous dragonfly found very locally in S Europe. Males eventually turn all-black, but often retain white appendages. Range has expanded to N less than other species that have colonized from Africa or Asia.

Identification: Small. In both sexes eyes brown above, blue-grey below; wings clear except for **small yellow-brown triangle** at base of hindwing (larger and paler in female); wings have rather sparse **dark** venation (pale on Black Pennant (*p. 332*)), wing-spot **large, amber-brown** (small and pale on Black Pennant); upper appendages usually **pale** (black on Black Pennant). ♂ Top of 'face' metallic **purple**; body and legs **black** when fully mature; abdomen slender. ♀ Body pale yellow with bold black markings; thorax has complex series of irregular black markings; abdomen has **pairs of pale marks**, decreasing in size and becoming triangular from base towards tip, between thick black line along top and black along lower side; top of S8–10 all-black; legs black, pale only at base. VARIATION: Immatures paler than female, extent of dark markings increasing with age; male retains series of pale 'wedges' on side of abdomen until fully mature; old females may become black, as can male appendages.

Behaviour: Males aggressive; typically flies a short distance if disturbed and perches inconspicuously low down in grassy vegetation.

Breeding habitat: Unshaded grassy ponds, lakes and swamps, occasionally ditches and slow-flowing streams; often at brackish coastal waters.

LOOK-ALIKES (in range)
Darters (*pp. 300–321*)
Indigo Dropwing (*p. 330*)
Black Pennant (*p. 332*)
Northern Banded Groundling
   (*p. 336*)

♀

♂ ♂+♀ small yellow-
brown patches at
base of hindwing

♀

♂i

♀o ♀ older individuals have
blacker markings and may
turn almost entirely black

♂

♂+♀ appendages
usually whitish

♀i

♂i

LC Northern Banded Groundling

Brachythemis impartita ×1·5

Locally common △

Lake, pond

J F M A M J J A S O N D

Overall length:	25–34 mm
Hindwing:	20–26 mm

Recognized as a separate species from Southern Banded Groundling *Brachythemis leucosticta*, which is widespread in Africa, as recently as 2009, this small ground-loving dragonfly flutters behind grazing animals, humans and even passing vehicles.

Identification: Small. Both sexes short and thickset; wing-spot **creamy-white with outer third or less dark** (uniformly pale on Black Pennant (*p. 332*), brown in Black Percher (*p. 334*)). ♂ Eyes dark brown above, blue-grey below; body, appendages and legs **black** when fully mature; wings have broad **dark brown band** between node and wing-spot. ♀ Eyes pale grey-brown with **three brown stripes** across top; body pale buff with black markings, irregular on thorax; abdomen has thick black **stripes along top and side**; appendages similar shape to those of male (shorter and widely spaced in other species in the family Libellulidae); wings usually clear, costa (from wing base to node) and antenodal cross-veins pale; legs yellowish with variable black markings. VARIATION: Immatures like female, bands across wings on males developing with maturity; male may retain pale appendages; female darkens with age, some developing faint wing bands.

Behaviour: Habitually perches on bare ground often away from water, flying low when disturbed. Territorial males perch on ground or low vegetation over or beside water. Huge numbers have been recorded around some reservoirs. Prone to wandering.

Breeding habitat: Open standing waters, especially large lakes with fluctuating water level and bare ground; occasionally slow-flowing rivers. Has spread into SW and C Mediterranean areas since the 1950s, and especially since the 1970s. Only vagrants to date in SE.

LOOK-ALIKES (in range)

Black Pennant (*p. 332*)
Black Percher (*p. 334*)
Banded Darter (*p. 320*)
♀ Small Skimmer (*p. 278*)

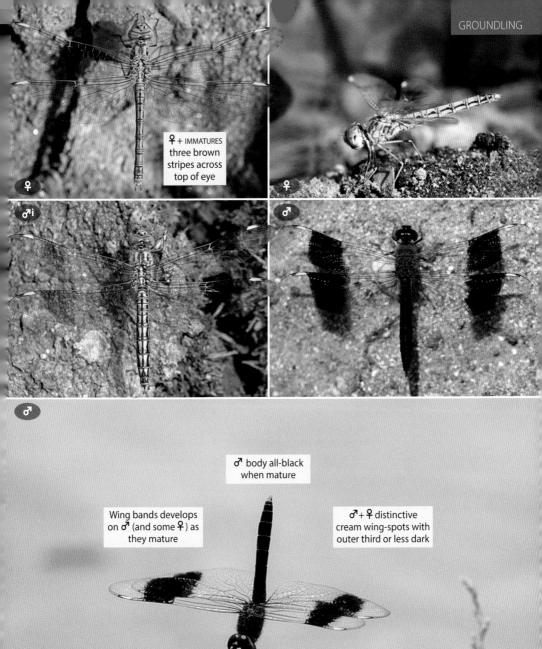

♀ + IMMATURES
three brown
stripes across
top of eye

♀

♀

♂i

♂

♂

♂ body all-black
when mature

Wing bands develops
on ♂ (and some ♀) as
they mature

♂ + ♀ distinctive
cream wing-spots with
outer third or less dark

GLIDERS

GENERA | *Tramea & Pantala*

FAMILY | **Libellulidae**

2 SPECIES

Medium-sized 'fliers' (mean length 47–50 mm) with broad-based hindwings; tapering abdomen, often held down at an angle to reduce exposure to the sun in order to avoid overheating; fly for long periods, often in groups; many undertake long migrations and are prone to vagrancy. Breed in fish-free pools, often temporary, including puddles left after rains.

Key features of gliders

◆ broad-based hindwings
◆ prolonged flight

Identify species by presence of dark patch in hindwing base | extent of black on abdomen pattern | perching position

Keyhole Glider

Tramea basilaris ×1·5

Wheeling Glider

Vagrant

Pond, open ground

J F M A M J J A S O N D

Overall length:	45–49 mm
Hindwing:	38–39 mm

The first European record was of a male (shown *below*) and female photographed in Italy at Linosa Island, between Sicily and Tunisia, on 21 October 2016. The closest known record to this is some 2,500 km away at Lake Chad!

Identification: Medium-sized. Similar to Wandering Glider (*p. 340*) but with diagnostic 'keyhole' at base of each hindwing easily seen in flight or when perched. In both sexes eyes red-brown above, blue-grey below; thorax unmarked brown; hindwings very **broad-based**, reaching almost halfway down abdomen; abdomen tapering, rather **short** and cylindrical with **S8–10 black** above; typically held down at an angle in flight; wing-spots reddish, **longer** on forewing than on hindwing. ♂ Abdomen **red**; hindwing base has large semicircular **brown patch with red veins around clear area at base** ('keyhole'), brown area surrounded by amber suffusion and yellow veins. ♀ Abdomen pale olive-brown; appendages same length as in male, angled inwards at tip (bent out in male); wing base pattern similar to male, but dark area fragmented, producing two irregular-shaped patches, and all veins yellow. VARIATION: Immatures as female, but yellower; in both sexes, abdomen may have black segment divisions, and dark wing base pattern variable.

Behaviour: High-flying, gliding flight similar to Wandering Glider; perches horizontally (although abdomen droops in hot conditions), sometimes high in trees at tip of twig. A long-distance migrant in Africa and Asia, its tendency to vagrancy is illustrated by several transatlantic records.

Breeding habitat: Breeds in reedy pools, but usually seen flying over open, grassy areas.

♂

LOOK-ALIKES (in range)
Wandering Glider (*p. 340*)

♂ + ♀ the 'keyhole' in dark wing base can be seen in flight

♂ + ♀ diagnostic clear 'keyholes' in wing base patches

♂ + ♀ broad-based, triangular hindwings enable sustained flight

♂ + ♀ abdomen usually unmarked apart from black on S8–10

NE Wandering Glider

Pantala flavescens ×1·5

Globe Skimmer

Rare ?

Pond, open ground

J F M A M J J A S O N D

Overall length:	45–55 mm
Hindwing:	38–42 mm

Amazing Migrations

Wandering Glider undertakes truly phenomenal mass migrations that follow seasonal rains in the tropics. Some fly over the Himalaya and cross the Indian Ocean before arriving to breed in East Africa. It is likely that their progeny then move into southern Africa, possibly with individuals from subsequent generations making the return journey, some of which may pass through Europe.

A common tropical migrant, considered to be the most widespread of all dragonfly species, having been recorded from all continents except Antarctica. It may undertake the longest migration of any insect.

Identification: Medium-sized, rather like large, long-winged chaser (*pp. 252–257*) or skimmer (*pp. 262–281*). Eyes red-brown above, blue-grey below; thorax tawny with unmarked yellowish side; hindwings very **broad-based** with **yellow patch** at base, reaching almost halfway down abdomen; abdomen tapering, rather **short** and cylindrical with **irregular dark central marking** that becomes blacker towards tip, typically held down at an angle in flight; wing-spots red-brown, **longer** on forewing than on hindwing. ♂ Abdomen **orange-red** on top, yellower on side; wingtips have small **amber patch**. ♀ Abdomen ochre; appendages longer than in male (length of S9+S10). VARIATION: Immatures have yellowish body; female abdomen becomes duller with age, and can be orange on top; extent of black on abdomen and wing markings variable: some only black on S8–9 and/or with clear wings.

Behaviour: Flies persistently and high with much gliding, abdomen characteristically angled down. Feeds in groups, often in swarms when migrating. Rarely settles, but when does so hangs at steep angle or vertically.

Breeding habitat: Small, warm, often temporary, pools (including puddles left after rain) and sometimes slow-flowing waters, usually with little vegetation, where life-cycle completed in 1–2 months; tolerates brackish water. Migrants often far from water or over sea, where may be attracted to ships' lights at night. Most records from SE in late summer or autumn, others possibly ship-assisted. Regular Cyprus, where breeding confirmed, and Linosa, Italy.

LOOK-ALIKES (in range)

Darters (*pp. 300–321*)
♀ chasers (*pp. 252–257*)
♀ skimmers (*pp. 262–281*)
Keyhole Glider (*p. 338*)

♂

♀

♂+♀ abdomen tip
angled down in flight

♀

♂+♀ tapering,
cylindrical abdomen has
dark marking along top

♂+♀ broad-based,
triangular hindwings
enable sustained flight

♂

Conservation and legislation

A significant number of the dragonfly species in Europe are rare or highly localized. As a consequence, many are of conservation concern and those that are particularly at risk are afforded legislative protection, including through requirements to conserve their habitats. This legislation is informed by evaluations of the status of all regular breeding species. The means by which species are assessed and the relevant legal measures are summarized in this section.

IUCN Red List Status

The International Union for Conservation of Nature (IUCN) has established global criteria for assessing the conservation status of species. These criteria include the rate of decline, population size, area of geographic distribution and degree of population and distribution fragmentation. Based on an analysis of these criteria, species are assigned to one of the following categories: Extinct (EX), Critically Endangered (CR), Endangered (EN), Vulnerable (VU), Near Threatened (NT), Least Concern (LC), Data Deficient (DD) or Not Evaluated (NE). An assessment of European Red List status of the breeding dragonfly species was published in the *European Red List of Dragonflies* in 2010 (see *below* for full reference); the codes from this assessment are included in Conservation Status boxes next to the species' names in the main species accounts, and in the *Checklist of species* (see *opposite*). Nineteen species were assessed as threatened: two as Critically Endangered, four as Endangered and 13 as Vulnerable.

The current Red List category for each species is indicated in the main species accounts and summarized in the *Checklist of species* (*opposite*).

International and European Union (EU) Legislation

Sixteen species are listed in Annex II and/or Annex IV of the EU Habitats Directive and in Appendix II of the Council for Europe Bern Convention on the Conservation of European Wildlife and Natural Habitats. As a consequence, signatories are required to take action to protect these species and their habitats. The 19 species placed in the Red List threat categories are also listed in Annexes II and IV of the Habitats Directive, together with Sedgling *Nehalennia speciosa*, which is assessed as Vulnerable within the EU. Fourteen of the 19 species are also included in Appendix II of the Bern Convention, three of which are also in the Red List threatened categories.

Protected species are highlighted with a red box in the main species accounts and shaded in pink in the *Checklist of species* (*opposite*).

Reference

Kalkman, V.J., Boudot, J.-P., Bernard, R., Conze, K.-J., De Knijf, G., Dyatlova, E., Ferreira, S., Jović, M., Ott, J., Riservato, E. & Sahlén, G. 2010. *European Red List of Dragonflies*. Luxembourg: Publications Office of the European Union (ec.europa.eu/environment/nature/conservation/species/redlist/downloads/European_dragonflies.pdf).

Checklist of species

This table is a complete list of European Odonata, based on the ***World Odonata List*** (Schorr & Paulson, 2020: http://bit.ly/3c2bM7e). It includes the authority responsible for describing each genus and species. Genera are listed alphabetically within families and species are listed alphabetically within genera. For each breeding species, the IUCN European Red List category is shown, together with a summary of the overall population trend; the 13 species endemic to Europe are also indicated. Species afforded special protection under the EU Habitats Directive and the Bern Convention are highlighted with pink shading (see *opposite* for further details of the relevant legislation). The final two columns provide an opportunity to record what you have seen, and a reference to the main account for each species.

E Endemic to Europe

Red List category (RL)
CR Critically Endangered
EN Endangered
VU Vulnerable
NT Near Threatened
LC Least Concern
DD Data Deficient
NE Not Evaluated

Population trend (PT)
= Stable
▲ Increasing
▼ Declining
? Unknown

Specially protected

E	RL	PT	SPECIES	English name	✓	Page
			SUBORDER **Zygoptera**	**Damselflies**		
			FAMILY **Lestidae**			
			GENUS ***Chalcolestes*** Kennedy, 1920			
	LC	?	*Chalcolestes parvidens* (Artobolevskii, 1929)	**Eastern Willow Spreadwing**		32
	LC	=	*Chalcolestes viridis* (Vander Linden, 1825)	**Western Willow Spreadwing**		30
			GENUS ***Lestes*** Leach, 1815			
	LC	=	*Lestes barbarus* (Fabricius, 1798)	**Migrant Spreadwing**		26
	LC	=	*Lestes dryas* Kirby, 1890	**Robust Spreadwing**		22
	VU	▼	*Lestes macrostigma* (Eversmann, 1836)	**Dark Spreadwing**		28
	LC	=	*Lestes sponsa* (Hansemann, 1823)	**Common Spreadwing**		20
	LC	=	*Lestes virens* (Charpentier, 1825)	**Small Spreadwing**		24
			GENUS ***Sympecma*** Burmeister, 1839			
	LC	▲	*Sympecma fusca* (Vander Linden, 1820)	**Common Winter Damsel**		34
	LC	▼	*Sympecma paedisca* (Brauer, 1877)	**Siberian Winter Damsel**		36
			FAMILY **Calopterygidae**			
			GENUS ***Calopteryx*** Leach, 1815			
	LC	=	*Calopteryx haemorrhoidalis* (Vander Linden, 1825)	**Copper Demoiselle**		46
	LC	=	*Calopteryx splendens* (Harris, 1780)	**Banded Demoiselle**		42
	LC	=	*Calopteryx virgo* (Linnaeus, 1758)	**Beautiful Demoiselle**		40
	LC	=	*Calopteryx xanthostoma* (Charpentier, 1825)	**Western Demoiselle**		44
			FAMILY **Euphaeidae**			
			GENUS ***Epallage*** Charpentier, 1840			
	NT	▼	*Epallage fatime* (Charpentier, 1840)	**Odalisque**		48

E	RL	PT	SPECIES	English name	✓	Page
			FAMILY **Platycnemididae**			
			GENUS *Platycnemis* Burmeister, 1839			
E	LC	=	*Platycnemis acutipennis* Selys, 1841	Orange Featherleg		58
E	LC	=	*Platycnemis latipes* Rambur, 1842	White Featherleg		60
	LC	=	*Platycnemis pennipes* (Pallas, 1771)	Blue Featherleg		62
			FAMILY **Coenagrionidae**			
			GENUS *Ceriagrion* Selys, 1876			
	CR	▼	*Ceriagrion georgifreyi* Schmidt, 1953	Turkish Red Damsel		56
	LC	=	*Ceriagrion tenellum* (de Villers, 1789)	Small Red Damsel		54
			GENUS *Coenagrion* Kirby, 1890			
	NT	▼	*Coenagrion armatum* (Charpentier, 1840)	Dark Bluet		82
	NT	▼	*Coenagrion caerulescens* (Fonscolombe, 1838)	Mediterranean Bluet		90
	LC	?	*Coenagrion hastulatum* (Charpentier, 1825)	Spearhead Bluet		74
	VU	=	*Coenagrion hylas* (Trybom, 1889)	Siberian Bluet		78
E	VU	?	*Coenagrion intermedium* Lohmann, 1990	Cretan Bluet		92
	LC	=	*Coenagrion johanssoni* (Wallengren, 1894)	Arctic Bluet		76
	LC	?	*Coenagrion lunulatum* (Charpentier, 1840)	Crescent Bluet		80
	NT	▼	*Coenagrion mercuriale* (Charpentier, 1840)	Mercury Bluet		84
	NT	▼	*Coenagrion ornatum* (Selys, 1850)	Ornate Bluet		86
	LC	=	*Coenagrion puella* (Linnaeus, 1758)	Azure Bluet		70
	LC	=	*Coenagrion pulchellum* (Vander Linden, 1825)	Variable Bluet		72
	LC	=	*Coenagrion scitulum* (Rambur, 1842)	Dainty Bluet		88
			GENUS *Enallagma* Charpentier, 1840			
	LC	=	*Enallagma cyathigerum* (Charpentier, 1840)	Common Bluet		68
			GENUS *Erythromma* Charpentier, 1840			
	LC	▲	*Erythromma lindenii* (Selys, 1840)	Blue-eye		96
	LC	=	*Erythromma najas* (Hansemann, 1823)	Large Redeye		98
	LC	▲	*Erythromma viridulum* (Charpentier, 1840)	Small Redeye		100
			GENUS *Ischnura* Charpentier, 1840			
	LC	=	*Ischnura elegans* (Vander Linden, 1820)	Common Bluetail		106
	VU	?	*Ischnura fountaineae* Morton, 1905	Oasis Bluetail		116
E	LC	=	*Ischnura genei* (Rambur, 1842)	Island Bluetail		110
	LC	=	*Ischnura graellsii* (Rambur, 1842)	Iberian Bluetail		108
	VU	▼	*Ischnura hastata* (Say, 1839)	Citrine Forktail		118
	NE	?	*Ischnura intermedia* Dumont, 1974	Persian Bluetail		104
	LC	=	*Ischnura pumilio* (Charpentier, 1825)	Small Bluetail		102
	LC	=	*Ischnura saharensis* Aguesse, 1958	Sahara Bluetail		112
	NE	?	*Ischnura senegalensis* (Rambur, 1842)	Marsh Bluetail		114
			GENUS *Nehalennia* Selys 1850			
	NT	▼	*Nehalennia speciosa* (Charpentier, 1840)	Sedgling		120
			GENUS *Pyrrhosoma* Charpentier, 1840			
E	CR	▼	*Pyrrhosoma elisabethae* Schmidt, 1948	Greek Red Damsel		52
	LC	=	*Pyrrhosoma nymphula* (Sulzer, 1776)	Large Red Damsel		50

E	RL	PT	SPECIES	English name	✓	Page
			SUBORDER **Anisoptera**	Dragonflies		
			FAMILY **Aeshnidae**			
			GENUS *Aeshna* Fabricius, 1775			
	LC	▲	*Aeshna affinis* Vander Linden, 1820	Blue-eyed Hawker		140
	LC	▼	*Aeshna caerulea* (Ström, 1783)	Azure Hawker		146
	NT	=	*Aeshna crenata* Hagen, 1856	Siberian Hawker		152
	LC	=	*Aeshna cyanea* (Müller, 1764)	Blue Hawker		142
	LC	=	*Aeshna grandis* (Linnaeus, 1758)	Brown Hawker		156
	LC	=	*Aeshna isoceles* (Müller, 1767)	Green-eyed Hawker		158
	LC	=	*Aeshna juncea* (Linnaeus, 1758)	Moorland Hawker		148
	LC	▲	*Aeshna mixta* Latreille, 1805	Migrant Hawker		138
	LC	=	*Aeshna serrata* Hagen, 1856	Baltic Hawker		154
	LC	▼	*Aeshna subarctica* Walker, 1908	Bog Hawker		150
	NT	▼	*Aeshna viridis* Eversmann, 1836	Green Hawker		144
			GENUS *Anax* Leach, 1815			
	LC	▲	*Anax ephippiger* (Burmeister, 1839)	Vagrant Emperor		172
	VU	?	*Anax immaculifrons* Rambur, 1842	Magnificent Emperor		174
	LC	▲	*Anax imperator* Leach, 1815	Blue Emperor		166
			Anax junius (Drury, 1773)	Common Green Darner		168
	LC	▲	*Anax parthenope* (Selys, 1839)	Lesser Emperor		170
			GENUS *Boyeria* McLachlan, 1895			
E	EN	▼	*Boyeria cretensis* Peters, 1991	Cretan Spectre		164
	LC	=	*Boyeria irene* (Fonscolombe, 1838)	Western Spectre		162
			GENUS *Brachytron* Evans, 1845			
	LC	=	*Brachytron pratense* (Müller, 1764)	Hairy Hawker		136
			GENUS *Caliaeschna* Selys, 1883			
	NT	▼	*Caliaeschna microstigma* (Schneider, 1845)	Eastern Spectre		160
			FAMILY **Gomphidae**			
			GENUS *Gomphus* Leach, 1815			
E	NT	▼	*Gomphus graslinii* Rambur, 1842	Pronged Clubtail		184
	LC	=	*Gomphus pulchellus* Selys, 1840	Western Clubtail		186
	NT	?	*Gomphus schneiderii* Selys, 1850	Turkish Clubtail		180
	LC	▼	*Gomphus simillimus* Selys, 1840	Yellow Clubtail		182
	LC	=	*Gomphus vulgatissimus* (Linnaeus, 1758)	Common Clubtail		178
			GENUS *Lindenia* de Haan, 1826			
	VU	▼	*Lindenia tetraphylla* (Vander Linden, 1825)	Bladetail		206
			GENUS *Onychogomphus* Selys, 1854			
	NE	?	*Onychogomphus assimilis* (Schneider, 1845)	Dark Pincertail		202
	EN	▼	*Onychogomphus costae* Selys, 1885	Faded Pincertail		198
	NE	?	*Onychogomphus flexuosus* (Schneider, 1845)	Waved Pincertail		200
*	LC	=	*Onychogomphus forcipatus* (Linnaeus, 1758)	Small Pincertail		194
	LC	=	*Onychogomphus uncatus* (Charpentier, 1840)	Large Pincertail		196
			GENUS *Ophiogomphus* Selys, 1854			
	LC	=	*Ophiogomphus cecilia* (Geoffroy in Fourcroy, 1785)	Green Snaketail		192

* subspecies *albotibialis* is **NT** and ▼

E	RL	PT	SPECIES	English name	✓	Page
			GENUS *Paragomphus* Cowley, 1934			
	LC	=	*Paragomphus genei* (Selys, 1841)	Green Hooktail		204
			GENUS *Stylurus* Needham, 1897			
	LC	▲	*Stylurus flavipes* (Charpentier, 1825)	River Clubtail		188
	DD	?	*Stylurus ubadschii* (Schmidt, 1953)	Syrian Clubtail		190
			FAMILY **Cordulegastridae**			
			GENUS *Cordulegaster* Leach, 1815			
E	NT	▼	*Cordulegaster bidentata* Selys, 1843	Sombre Goldenring		220
	LC	=	*Cordulegaster boltonii* (Donovan, 1807)	Common Goldenring		212
E	EN	▼	*Cordulegaster helladica* (Lohmann, 1993)	Greek Goldenring		222
E	NT	=	*Cordulegaster heros* Theischinger, 1979	Balkan Goldenring		216
	EN	?	*Cordulegaster insignis* Schneider, 1845	Blue-eyed Goldenring		224
	VU	=	*Cordulegaster picta* Selys, 1854	Turkish Goldenring		218
E	NT	▼	*Cordulegaster trinacriae* Waterston, 1976	Italian Goldenring		214
			incertae sedis [taxonomic relationship/position uncertain]			
			GENUS *Oxygastra* Selys, 1870			
	LC	=	*Oxygastra curtisii* (Dale, 1834)	Orange-spotted Emerald		248
			FAMILY **Macromiidae**			
			GENUS *Macromia* Rambur, 1842			
E	VU	▼	*Macromia splendens* (Pictet, 1843)	Splendid Cruiser		226
			FAMILY **Corduliidae**			
			GENUS *Cordulia* Leach, 1850			
	LC	=	*Cordulia aenea* (Linnaeus, 1758)	Downy Emerald		232
			GENUS *Corduliochlora* Marinov & Seidenbusch, 2007			
	VU	▼	*Corduliochlora borisi* Marinov, 2001	Bulgarian Emerald		234
			GENUS *Epitheca* Burmeister, 1839			
	LC	=	*Epitheca bimaculata* (Charpentier, 1825)	Eurasian Baskettail		250
			GENUS *Somatochlora* Selys, 1871			
	LC	?	*Somatochlora alpestris* (Selys, 1840)	Alpine Emerald		244
	LC	?	*Somatochlora arctica* (Zetterstedt, 1840)	Northern Emerald		242
	LC	=	*Somatochlora flavomaculata* (Vander Linden, 1825)	Yellow-spotted Emerald		240
	LC	?	*Somatochlora meridionalis* Nielsen, 1935	Balkan Emerald		238
	LC	=	*Somatochlora metallica* (Vander Linden, 1825)	Brilliant Emerald		236
	DD	?	*Somatochlora sahlbergi* Trybom, 1889	Treeline Emerald		246
			FAMILY **Libellulidae**			
			GENUS *Brachythemis* Brauer, 1868			
	LC	▲	*Brachythemis impartita* (Karsch, 1890)	Northern Banded Groundling		336
			GENUS *Crocothemis* Brauer, 1868			
	LC	▲	*Crocothemis erythraea* (Brullé, 1832)	Broad Scarlet		322
			GENUS *Diplacodes* Kirby, 1889			
	LC	=	*Diplacodes lefebvrii* (Rambur, 1842)	Black Percher		334
			GENUS *Leucorrhinia* Brittinger, 1850			
	LC	=	*Leucorrhinia albifrons* (Burmeister, 1839)	Dark Whiteface		290

E	RL	PT	SPECIES	English name	✓	Page
	LC	=	*Leucorrhinia caudalis* (Charpentier, 1840)	**Lilypad Whiteface**		*292*
	LC	=	*Leucorrhinia dubia* (Vander Linden, 1825)	**Small Whiteface**		*284*
	LC	▼	*Leucorrhinia pectoralis* (Charpentier, 1825)	**Yellow-spotted Whiteface**		*288*
	LC	▼	*Leucorrhinia rubicunda* (Linnaeus, 1758)	**Ruby Whiteface**		*286*
GENUS ***Libellula*** Linnaeus, 1758						
	LC	=	*Libellula depressa* Linnaeus, 1758	**Broad-bodied Chaser**		*254*
	LC	=	*Libellula fulva* Müller, 1764	**Blue Chaser**		*256*
	LC	=	*Libellula quadrimaculata* Linnaeus, 1758	**Four-spotted Chaser**		*252*
GENUS ***Orthetrum*** Newman, 1833						
	LC	▲	*Orthetrum albistylum* (Selys, 1848)	**White-tailed Skimmer**		*264*
	LC	▲	*Orthetrum brunneum* (Fonscolombe, 1837)	**Southern Skimmer**		*268*
	LC	=	*Orthetrum cancellatum* (Linnaeus, 1758)	**Black-tailed Skimmer**		*262*
	LC	=	*Orthetrum chrysostigma* (Burmeister, 1839)	**Epaulet Skimmer**		*274*
	LC	=	*Orthetrum coerulescens* (Fabricius, 1798)	**Keeled Skimmer**		*266*
	VU	▼	*Orthetrum nitidinerve* (Selys, 1841)	**Yellow-veined Skimmer**		*270*
			Orthetrum ransonnetii (Brauer, 1865)	**Desert Skimmer**		*272*
	LC	=	*Orthetrum sabina* (Drury, 1773)	**Slender Skimmer**		*280*
	LC	=	*Orthetrum taeniolatum* (Schneider, 1845)	**Small Skimmer**		*278*
	LC	=	*Orthetrum trinacria* (Selys, 1841)	**Long Skimmer**		*276*
GENUS ***Pantala*** Hagen, 1861						
	NE	?	*Pantala flavescens* (Fabricius, 1798)	**Wandering Glider**		*340*
GENUS ***Selysiothemis*** Ris, 1897						
	LC	▲	*Selysiothemis nigra* (Vander Linden, 1825)	**Black Pennant**		*332*
GENUS ***Sympetrum*** Newman, 1883						
	LC	=	*Sympetrum danae* (Sulzer, 1776)	**Black Darter**		*300*
	VU	▼	*Sympetrum depressiusculum* (Selys, 1841)	**Spotted Darter**		*314*
	LC	=	*Sympetrum flaveolum* (Linnaeus, 1758)	**Yellow-winged Darter**		*318*
	LC	▲	*Sympetrum fonscolombii* (Selys, 1840)	**Red-veined Darter**		*316*
	LC	▲	*Sympetrum meridionale* (Selys, 1841)	**Southern Darter**		*308*
E	LC	=	*Sympetrum nigrifemur* (Selys, 1884)	**Island Darter**		*306*
	LC	=	*Sympetrum pedemontanum* (Müller in Allioni, 1766)	**Banded Darter**		*320*
	LC	=	*Sympetrum sanguineum* (Müller, 1764)	**Ruddy Darter**		*312*
	LC	=	*Sympetrum sinaiticum* Dumont, 1977	**Desert Darter**		*310*
	LC	=	*Sympetrum striolatum* (Charpentier, 1840)	**Common Darter**		*302*
	LC	=	*Sympetrum vulgatum* (Linnaeus, 1758)	**Moustached Darter**		*304*
GENUS ***Tramea*** Hagen, 1861						
			Tramea basilaris (Palisot de Beauvois, 1817)	**Keyhole Glider**		*338*
GENUS ***Trithemis*** Brauer, 1868						
	LC	▲	*Trithemis annulata* (Palisot de Beauvois, 1807)	**Violet Dropwing**		*328*
	NE	?	*Trithemis arteriosa* (Burmeister, 1839)	**Red-veined Dropwing**		*326*
	LC	=	*Trithemis festiva* (Rambur, 1842)	**Indigo Dropwing**		*330*
	NE	▲	*Trithemis kirbyi* (Selys, 1891)	**Orange-winged Dropwing**		*324*
GENUS ***Zygonyx*** Selys in Hagen, 1867						
	VU	▼	*Zygonyx torridus* (Kirby, 1889)	**Ringed Cascader**		*228*

Further information

The books listed below are all currently in print and are either identification guides that cover the dragonflies of Europe or provide a useful general introduction to Odonata. The IT resources listed also have Europe-wide relevance.

Books

Askew, R. R. 1988 (2004). *The Dragonflies of Europe*. Harley Books.

Boudot, J.-P. & Kalkman, V. J. (EDS.). 2015. *Atlas of the European dragonflies and damselflies*. KNNV publishing, the Netherlands.

Brooks, S. 2003. *Dragonflies*. Natural History Museum.

Chandler, D & Cham, S. 2013. *Dragonfly*. New Holland Publishers.

Corbet, P. S. 1999 (2004, 2014). *Dragonflies: Behaviour and Ecology of Odonata*. E J Brill.

Dijkstra, K-D. B. & Lewington, R. 2006. *Field Guide to the Dragonflies of Britain and Europe*. British Wildlife Publishing. [Spanish edition: *Guia de Campo de las Libélulas de España y de Europa*. Ediciones Omega. German edition: *Libellen Europas: Der Bestimmungsführer*. Verlag Paul Haupt.]

Galliani, C., Scherini, R. & Piglia, A. 2017. *Dragonflies and Damselflies of Europe: A scientific approach to the identification of European Odonata without capture*. WBA Handbook No. 7.

Paulson, D. 2019. *Dragonflies & Damselfies: A natural history*. Ivy Press.

Wildermuth, H. & Martens, A. 2019. *Die Libellen Europas: Alle Arten von den Azoren bis zum Ural im Porträt*. Quelle & Meyer Verlag GmbH & Co., Wiebelsheim.

IT resources

DragonflyPix – Excellent images and online identification guide (dragonflypix.com)

Observation International – international wildlife record submission (observation.org)

An example of the stunning images taken by Fons Peels, many of which feature in this book; more can be found on his website DragonflyPix.com.

BLUE-EYED HAWKER
Aeshna affinis

Acknowledgements & photographic credits

Many people have contributed to the production of this book and our sincere thanks go to all. It is our intention that everyone who has contributed is named in this section, but if we have inadvertently missed anyone we can only apologize. Despite the contributions of others, we hold full responsibility for any errors or omissions.

The Princeton **WILD***Guides* series of field guides covering Britain's and Europe's wildlife is the brainchild of **Rob Still**, who is responsible for their design and production. Rob's unrivalled – yet constantly developing – skills in computer graphics have been used to wonderful effect in designing this book, and he has also produced the many illustrations of key features that are included throughout. This book would be nowhere near as attractive and accessible without Rob's contribution and we recognise with grateful thanks the countless hours he has spent seeing it to fruition.

The key feature of this book is undoubtedly the remarkable photographs that grace the pages. Without the contribution of the 81 photographers from across Europe and farther afield who kindly allowed us to use their work, its production would not have been possible. To all we owe a great debt of gratitude. Particular thanks must go to **Fons Peels**, whose remarkable images are showcased throughout. Fons is the man behind DragonfyPix.com, which provides a treasure trove of information as well as many images of males and females of almost every species of damselfly and dragonfly included in this book. His willingness to contribute to the book and his patience during the image selection process is very gratefully acknowledged. **Christophe Brochard** (cbrochard.com) and **Erland Refling Nielsen** (flickr.com/photos/23985726@N05) generously supplied a large number of excellent images. Other major contributors were **Paul Cools**, **Wil Leurs**, **Pablo Martinez-Darve Sanz** and **Fazal Sardar**.

A special mention goes to **Rachel** and **Anya Still**, **Gill Swash** and **Sue Smallshire** for their invaluable contributions behind the scenes, and to **Brian Clews** for his commitment and dedication in sourcing images, and his keen eye in checking the final draft. **Robert Kirk**, Publisher, Field Guides & Natural History, at Princeton University Press, has encouraged and helped us throughout this project.

We would also like to express our gratitude to **Jean-Pierre Boudot**, **Francisco Jesus Cano**, **K-D. B. Dijkstra**, **Vincent Kalkman**, **Yordan Kutsarov**, **Milen Marinov**, **Adrian Parr** and **Nick Ransdale** (nick-ransdale.com) for their help, guidance and support during the production process.

Photographic credits

All 1,417 photographs included in the book are listed on the following pages. The contributing photographers are listed overleaf (*p. 350*), together with the initials that are used to identify each of their images. Images sourced via the photographic agencies Agami (agami.nl) and Alamy (alamy.com) are indicated with the agency's name after the photographer's initials. A number of images are reproduced under the terms of the Creative Commons Attribution-ShareAlike 2.0 Generic license, Creative Commons Attribution-ShareAlike 3.0 Unported license or the Creative Commons Attribution-ShareAlike 4.0 International license; these are indicated by "/CC" after the photographer's initials in the list, and where appropriate the relevant license is referenced.

Contributing photographers

Mike Averill [MA]; **Miloš Balla** [MB]; **Jean-Pierre Boudot** [JPB]; **Jaap Bouwman** [JBou]; **John Bowler** [JBow]; **Thomas Bresson** (flickr.com/photos/computerhotline) [TB/CC]; **Paul Brock** [PB]; **Christophe Brochard** (cbrochard.com) [CB]; **Vicente Camacho-Lozano** (flickr. com/photos/27663496@N08) [VCL]; **Francisco J. Cano-Villegas** [FJCV]; **Tim Caroen** (flickr. com/photos/tim-pc) [TC]; **Steve Cham** [SCh]; **Andrew Chambers** [AC]; **Ivan Chiandetti** [IC]; **Jay Coleman** [JC/CC]; **Paul Cools** (pbase.com/paulcoolsphotography) [PC]; **Steve Covey** [SCo]; **John & Carol Curd** (odonata.org.uk) [J&CC]; **Theo Douma** (Agami.nl) [TD/ Agami]; **Christian Ferrer** [CFe/CC]; **Christian Fischer** (CC BY-SA 3.0 (creativecommons. org/licenses/by-sa/3.0)], via Wikimedia Commons) [CFi/CC]; **Miikka Friman** [MF]; **Robert Geerts** (odonata.eu) [RG]; **Geoff Gowlett** [GG]; **Benoit Guillon** (meslibellules.fr) [BG]; **Marc Guyt** (Agami.nl) [MG/Agami]; **Marc Heath** [MH]; **Gail Hampshire** (flickr.com/photos/ gails_pictures) [GH/CC]; **Paul Harrison** [PH]; **Bas Haasnoot** (Agami.nl) [BH/Agami]; **Annie Irving** (earthstar.blog) [AI]; **Jim Johnson** (odonata.bogfoot.net) [JJ]; **Sami Karjalainen** (korento.net) [SK]; **Thomas Kirchen** (makro-tom.de) [TK]; **Miroslaw Krol** (flickr.com/ photos/mirekkrol) [MK]; **A. N. Suresh Kumar** (flickr.com/photos/ansk) [ANSK]; **Yordan Kutsarov** [YK]; **Wil Leurs** (Agami.nl) [WL/Agami]; **Bert Logtmeijer** [BL]; **Roy Lowry** (flickr. com/photos/99817330@N02) [RL]; **René Manger** [RM]; **Gilles San Martin** [GSM/CC]; **Pablo Martinez-Darve Sanz** (flickr.com/photos/blezsp) [PMDS]; **Andy McGeeney** [AM]; **Nicolas Mézière** [NM]; **Iñaki Mezquita** [IM]; **Simon Mitchell** [SM]; **Erland Refling Nielsen** (flickr.com/photos/23985726@N05) [ERN]; **Simon Oliver** (flickr.com/photos/126688171@ N06) [SO]; **Dennis Paulson** [DP]; **Cecile Peels** [CP]; **Fons Peels** (DragonflyPix.com) [FP]; **Damian Keith Pinguey** (flickr.com/photos/rezamink) [DKP]; **Ian Preston** (flickr.com/ photos/9750464@N02) [IP/CC]; **Alastair Rae** (flickr.com/photos/merula) [AR]; **Paul Ritchie** (hampshiredragonflies.co.uk) [PR]; **Thibaut Rivière** (flickr.com/photos/95289326@N03) [TR]; **Gino Roncaglia** (flickr.com/photos/roncaglia) [GR/CC]; **Hermes Sarapuu** [HS]; **Fazal Sardar** (Agami.nl) [FS/Agami]; **Charles J. Sharp** (flickr.com/photos/93882360@N07) [CJS]; **Dave Smallshire** [DSm]; **Sue Smallshire** [SS]; **Walter Soestbergen** (Agami.nl) [WS/ Agami]; **Dave Sparrow** [DSp]; **Andy & Gill Swash** (WorldWildlifeImages.com) [A&GS]; **Terres de France (Resort & hotel nature)** [TdF/CC]; **Tihomir Stefanov** [TS]; **Pam Taylor** [PT]; **Antonio E. Tetares Ventura** (flickr.com/photos/aetven) [AETV]; **Jukka Toivanen** (flickr. com/photos/155315483@N07) [JT]; **Stefano Trucco** [ST]; **Paulo Valdivieso** (flickr.com/ photos/p_valdivieso) [PV/CC]; **Johannes van Donge** (diginature.nl) [JvD]; **Frank Vassen** (flickr.com/photos/42244964@N03) [FV/CC]; **Michele Viganò** [MV]; **Barry Warrington** [BW]; **Ray Wilson/Alamy Stock Photo** (alamy.com) [RW/Alamy] and **Paul Winter** [PW].

The following codes are used where necessary for clarity: M = male; F = female.

Images in the species account plates are **coded according to their position in the 'grid':**
A = top 'row', *B* = 2nd 'row' down, *C* = 3rd 'row' down, *etc.*; *1* = 1st 'column', *2* = 2nd 'column', *etc.*

Cover: **Beautiful Demoiselle** [A&GS].

Title page: **Violet Dropwing** [ERN].

INTRODUCTION p4: **Common Darter** [DSm]; **Blue-eyed Hawker** [MH]; **Violet Dropwing** [DSm]. p5: **Bladetail** [DSm]; **Large Red Damsel** [MH]; **Banded Darter** [DSm]; **Hairy Hawker** [DSm]; **Blue Featherleg** [DSm]. p8: **Spreadwing** [FP]; **Winter damsel** [BG]; **Demoiselle** [A&GS]; **Odalisque** [PC]. p9: **Red damsel** [A&GS]; **Featherleg** [A&GS]; **Bluet** [DSm]; **Brighteye** [FS/Agami]; **Bluetail** [TB/CC];

Sedgling [ERN]. **p10: Hawker** [PR]; **Emperor** [A&GS]; **Clubtail** [JT]; **Pincertail** [DSm]; **Goldenring** [FP]. **p11: Emerald** [PR]; **Chaser** [A&GS]; **Skimmer** [A&GS]; **Whiteface** [CFi/CC]; **Percher** [PW]; **Glider** [A&GS]. **p12: Owl-fly** [A&GS]; **Antlion** [DSm]; **Mayfly** [DSm]; **Lacewing** [PB]. **p13: Common Spreadwing** M (abdomen tip) [DSm], F (abdomen tip) [JT]; **Common Bluet** [TK]; **Mercury Bluet** (head/thorax) [A&GS]; **Hairy Hawker** (above) [JT], (side) [TK]; **Common Clubtail** (wings) [JT]; **Black Darter** (abdomen tip) [A&GS]; **Brown Hawker** F (abdomen tip) (top) [DSm], F (abdomen tip) (side) [JvD], M (abdomen tip) (top) [DSm], M (abdomen tip) (side) [TK].

INTRODUCTION TO DAMSELFLIES (ZYGOPTERA) p16: Common Spreadwing [ERN]; **Common Winter Damsel** [TS]; **Beautiful Demoiselle** [DSm]; **Odalisque** [PC]. **p17: Large Red Damsel** [A&GS]; **Blue Featherleg** [A&GS]; **Azure Bluet** [FP]; **Large Redeye** [ERN]; **Common Bluetail** [TB/CC]; **Sedgling** [ERN].

SPREADWINGS AND WINTER DAMSELS p18: Common Spreadwing [DSm]; **Western Willow Spreadwing** [ERN]; **spreadwing scars** [DSm]. **p19: THORAX SIDE: Common Spreadwing** [CB]; **Robust Spreadwing** [FP]; **Small Spreadwing** [WL/Agami]; **Migrant Spreadwing** [FV/CC]; **Dark Spreadwing** [WL/Agami]; **Western Willow Spreadwing** [FP]; **Eastern Willow Spreadwing** [FP]. **WING-SPOT: Common Spreadwing** [A&GS]; **Robust Spreadwing** [FP]; **Small Spreadwing** [DSm]; **Migrant Spreadwing** [DSm]; **Dark Spreadwing** [BG]; **Western Willow Spreadwing** [FP]; **Eastern Willow Spreadwing** [DSm]. **MALE APPENDAGES: Common Spreadwing** [BG]; **Robust Spreadwing** [BG]; **Small Spreadwing** [BG]; **Migrant Spreadwing** [BG]; **Dark Spreadwing** [FP]; **Western Willow Spreadwing** [DSm]; **Eastern Willow Spreadwing** [DSm]. **OVIPOSITOR: Common Spreadwing** [CB]; **Robust Spreadwing** [CB]; **Small Spreadwing** [ERN]; **Migrant Spreadwing** [FV/CC]; **Dark Spreadwing** [WL/Agami]; **Western Willow Spreadwing** [DSm]; **Eastern Willow Spreadwing** [DSm]. **Common Winter Damsel** [TS]. **p20: Common Spreadwing** habitat [CB]. **p21: Common Spreadwing** *A1* [ERN], *A/B2* [ERN], *B1* [CB], *C1* [CB], *C/D2* [ERN], *D1* [ERN]. **p22: Robust Spreadwing** habitat [CB]. **p23: Robust Spreadwing** *A1* [CB], *A2* [ERN], *B1* [FP], *B2* [FP], *C1* [FP], *C2* [FP]. **p24: Small Spreadwing** habitat [DSm]. **p25: Small Spreadwing** *A1* [ERN], *A2* [FP], *B1* [ERN], *B2* [ERN], *C1* [ERN], *C2* [ERN], inset (head) [DSm]. **p26: Migrant Spreadwing** habitat [CB]. **p27: Migrant Spreadwing** *A1* [FV/CC], *A2* [FS/Agami], *B1* [FP], *B2* [FP], inset (head) [DSm]. **p28: Dark Spreadwing** habitat [DSm]. **p27: Dark Spreadwing** *A1* [DSm], *A/B2* [DSm], *B1* [WL/Agami], *B/C2* [A&GS], *C1* [WL/Agami]. **p30: Western Willow Spreadwing** habitat [SCh]. **p31: Western Willow Spreadwing** *A1* [FP], *A2* [DSm], *B1* [FP], *B2* [FP]. **p32: Eastern Willow Spreadwing** habitat [CB]. **p33: Eastern Willow Spreadwing** all (4) [FP]. **p34: Common Winter Damsel** habitat [CB]. **p35: Common Winter Damsel** *A1* [ERN], *A2* [ERN], *A3* [TS], *B1* [BG], *B2* [WL/Agami], *C1* [FP], *C2* [BG]. **p36: Siberian Winter Damsel** habitat [CB]. **p37: Siberian Winter Damsel** *A1* [ERN], *A2* [CB], *B1* [CB], *B2* [ERN], *C1* [WL/Agami], *C2* [ERN].

DEMOISELLES p38: Banded Demoiselle M [TD/Agami], F (wing-spot) [DSm]; **Copper Demoiselle** (abdomen tip) [DSm]. **p39: Beautiful Demoiselle** *virgo* M [ERN], F [ERN], *festiva* M [DSm], F [CB]; **Banded Demoiselle** *splendens* M [A&GS], F [IP/CC], *cretensis* [CB], *balcanica* [FP]; **Western Demoiselle** M [BG], F [ERN]; **Copper Demoiselle** M (top) [A&GS], M (bottom) [GG], F (top) [FP], F (bottom) [A&GS]. **p40: Beautiful Demoiselle** habitat [DSm]. **p41: Beautiful Demoiselle** *A1* [A&GS], *B1* [ERN], *B2* [DSm], *C1* [ERN], *C2* [CB]. **p42: Banded Demoiselle** habitat [CB]. **p43: Banded Demoiselle** *A1* [IP/CC], *A2* [FP], *B1* [A&GS], *B2* [DSm], *C1* [DSm], *C2* [CB]. **p44: Western Demoiselle** habitat [CB]. **p45: Western Demoiselle** *A1* [BG], *B1* [ERN], *C1* [ERN], inset (abdomen tip) [ERN]. **p46: Copper Demoiselle** habitat [DSm]. **p47: Copper Demoiselle** *A1* [A&GS], *A2* [DSm], *B1* [GG], *C1* [DSm], *C2* [FP].

ODALISQUE p48: Odalisque habitat [CB], inset (pair) [DSm]. **p49: Odalisque** *A1* [PC], *B1* [DSm], *B2* [DSm], *C1* [A&GS], inset (dark F) [DSm].

RED DAMSELS p50: Large Red Damsel habitat [CB]. **p51: Large Red Damsel** *A/C1* [A&GS], *A2* [A&GS], *B2* [SCo], *C2* [FS/Agami], *D1* [RL], *D2* [A&GS], *D3* [DSm]. **p52: Greek Red Damsel** habitat [CB], inset (pronotum) [CB]; **Large Red Damsel** inset (pronotum) [DSm]. **p53: Greek Red Damsel** *A1* [CB], *A/B2* [CB], *B1* [FP], *C1* [FP], *C2* [FP], *D1* [CB]; **Large Red Damsel** inset (abdomen tip) [DSm]. **p54: Small Red Damsel** habitat [DSm]. **p55: Small Red Damsel** *A1* [A&GS], *A/C2* [A&GS], *B1* [ERN], *C1* [WL/Agami], *D1* [ERN], *D2* [A&GS]. **p56: Turkish Red Damsel** habitat [CB], inset (thorax) [FP]; **Small Red Damsel** inset (thorax) [ERN]. **p57: Turkish Red Damsel** *A1* [FP], *A2* [FP], *B1* [FP], *B2* [FP], *C1* [PC], *D1* [FP], *D2* [FP].

FEATHERLEGS p58: Orange Featherleg habitat [CB]. **p59: Orange Featherleg** *A1* [FP], *A/B2* [ERN], *B1* [FP], *C1* [CB], *C2* [FP], *D2* [DSm]. **p60: White Featherleg** habitat [CB], inset (legs) [FP]; **Blue Featherleg** inset (legs) [DSm]. **p61: White Featherleg** *A1* [FP], *A2* [FP], *B1* [FP], *B2* [A&GS], *C1* [FP], *C2* [FP], *D1* [WL/Agami], *D2* [BG]. **p62: Blue Featherleg** habitat [A&GS], inset (legs) [DSm]; **White Featherleg** inset (legs) [FP]. **p63: Blue Featherleg** *A1* [DSm], *B1* [FP], *B2* [FP], *C1* [BH/Agami], *C2* [FP], *D1* [MG/Agami], *D2* [DSm].

BLUETS **p64: Mercury Bluet** [A&GS], head (inset) [BG]; **Blue Featherleg** [A&GS]; **Large Redeye** M [PC]; **Common Bluetail** [DSm]. **p65: Common Bluet** [DSm]; **Azure Bluet** [DSm]; **Blue-eye** [DSm]; **Variable Bluet** [ERN]; **Mercury Bluet** [A&GS]; **Ornate Bluet** [DSm]; **Crescent Bluet** [ERN]; **Spearhead Bluet** [DSm]; **Arctic Bluet** [FP]; **Dark Bluet** [FP]; **Dainty Bluet** [FP]; **Siberian Bluet** [FP]; **Mediterranean Bluet** [FP]; **Cretan Bluet** [FP]. **p66: Common Bluet** (head/thorax) [FP]; **Azure Bluet** (head/thorax) [CB]. **p67: Common Bluet** [CJS]; **Azure Bluet** [BW]; **Blue-eye** [DSm]; **Variable Bluet** [ERN]; **Mercury Bluet** [FP]; **Ornate Bluet** [DSm]; **Crescent Bluet** [FP]; **Spearhead Bluet** [ERN]; **Arctic Bluet** [FP]; **Dark Bluet** [ERN]; **Dainty Bluet** [DSm]; **Siberian Bluet** [FP]; **Mediterranean Bluet** [FP]; **Cretan Bluet** [FP] **p68: Common Bluet** habitat [DSm]. **p69: Common Bluet** *A1* [FP], *A/B2* [DSm], *B1* [WL/Agami], *C1* [CJS], *C2* [FP], *D1* [FP], *D2* [DSm], inset (vulvar spine) [DSm]. **p70: Azure Bluet** habitat [CB]. **p71: Azure Bluet** *A1* [CB], *A2* [FP], *A3* [BW], *B1* [WS/Agami], *B2* [CB], *B3* [FP], *C1* [FP], *C2* [FP]. **p72: Variable Bluet** habitat [PT]. **p73: Variable Bluet** *A1* [ERN], *A2* [DSm], *B1* [ERN], *B2* [PC], *C1* [CB], *C2* [BG], *D1* [ERN], *D2* [ERN], *E1* [DSm], *E2* [ERN]. **p74: Spearhead Bluet** habitat [CB]. **p75: Spearhead Bluet** *A1* [ERN], *A2* [AR], *B1* [ERN], *B2* [ERN], *C1* [CB], *C2* [CB]. **p76: Arctic Bluet** habitat [CB]. **p77: Arctic Bluet** *A1* [ERN], *A2* [FP], *B1* [FP], *B2* [ERN], *C1* [FP], *C2* [FP]. **p78: Siberian Bluet** habitat [CB]. **p79: Siberian Bluet** *A1* [FP], *A2* [FP], *B1* [FP], *B2* [CB], *C1* [FP], *C2* [WL/Agami]. **p80: Crescent Bluet** habitat [CB]. **p81: Crescent Bluet** *A1* [FP], *A2* [FP], *B1* [FP], *B2* [ERN], *C1* [CB], *C2* [FS/Agami], *D1* [ERN]. **p82: Dark Bluet** habitat [CB]. **p83: Dark Bluet** *A1* [ERN], *A2* [ERN], *B1* [ERN], *B2* [ERN], *C1* [ERN], *C2* [CB]. **p84: Mercury Bluet** habitat [CB]. **p85: Mercury Bluet** *A1* [FP], *A2* [A&GS], *B1* [FP], *B2* [A&GS], *C1* [IM], *C2* [A&GS]. **p86: Ornate Bluet** habitat [CB]. **p87: Ornate Bluet** *A1* [FP], *A2* [DSm], *B1* [FP], *B2* [PC], *C1* [PC], *C2* [DSm]. **p88: Dainty Bluet** habitat [DSm]. **p89: Dainty Bluet** *A1* [FP], *A/B2* [DSm], *B1* [FS/Agami], *C/D1* [FS/Agami], *C2* [FP], *D2* [CB]. **p90: Mediterranean Bluet** habitat [CB]. **p91: Mediterranean Bluet** *A1* [FP], *A2* [FP], *B1* [FP], *B2* [FP], *C/D1* [FP], *C2* [FP], *D2* [FP]. **p92: Cretan Bluet** habitat [CB]. **p93: Cretan Bluet** *A1* [FP], *A2* [FP], *B1* [CB], *B2* [FP], *C1* [CB].

'BLUE-TAILED' DAMSELFLIES **p95: Blue-eye** [DSm]; **Large Redeye** [PC]; **Small Redeye** [DSm]; **Sedgling** [ERN]; **Common Bluetail** [GG]; **Iberian Bluetail** [FP]; **Island Bluetail** [FP]; **Oasis Bluetail** [FP]; **Sahara Bluetail** [FP]; **Marsh Bluetail** [FP]; **Small Bluetail** [ERN]; **Persian Bluetail** [FP]. **p96: Blue-eye** habitat [CB]; inset (M) [DSm]. **p97: Blue-eye** *A1* [FP], *A2* [ERN], *B1* [ST], *B2* [ERN], *C1* [FP], *C2* [WL/Agami], *D1* [FP]. **p98: Large Redeye** habitat [A&GS], inset (abdomen tip side) [FS/Agami], inset (abdomen tip above) [FP]; **Small Redeye** inset (abdomen tip side) [CB], inset (abdomen tip above) [DSm]. **p99: Large Redeye** *A/B1* [ERN], *A2* [FP], *B2* [FS/Agami], *C1* [ERN], *C2* [ERN], *D1* [CB], *D2* [ERN]. **p100: Small Redeye** habitat [CB], inset (abdomen tip side) [CB], inset (abdomen tip above) [DSm]; **Large Redeye** inset (abdomen tip side) [FS/Agami], inset (abdomen tip above) [FP]. **p101: Small Redeye** *A1* [DSm], *A2* [A&GS], *B1* [DSm], *B2* [DSm], *C1* [FP], *C2* [CB]. **p102: Small Bluetail** habitat [DSm]. **p103: Small Bluetail** *A1* [FP], *A2* [DSm], *B1* [WS/Agami], *B2* [ERN], *C1* [DSm], *C2* [ERN], *D1* [FS/Agami], *D2* [ERN]. **p104: Persian Bluetail** habitat [DSp]. **p105: Persian Bluetail** *A1* [FP], *A2* [FP], *B1* [FP], *B2* [FP], *C1* [FP], *C/D2* [FP], *D1* [FP]. **p106: Common Bluetail** habitat [PT]. **p107: Common Bluetail** *A1* [ERN], *A2* [GG], *B1* [ERN], *B2* [A&GS], *C1* [A&GS], *C2* [A&GS], *D1* [ERN], *D/E2* [ERN], *E1* [ERN]. **p108: Iberian Bluetail** habitat [CB]. **p109: Iberian Bluetail** *A1* [DSm], *A2* [DSm], *B1* [FP], *B2* [FP], *C1* [FP], *C2* [FP], *D1* [FP], *D2* [ERN], *E1* [FP], *E2* [ERN]. **p110: Island Bluetail** habitat [CB]. **p111: Island Bluetail** *A1* [FP], *A2* [FP], *B1* [FP], *B2* [FP], *C1* [FP], *C2* [FP], *D1* [CB]. **p112: Sahara Bluetail** habitat [JBow]. **p113: Sahara Bluetail** *A/B1* [FP], *A2* [FP], *B2* [PMDS], *C1* [PMDS], *C2* [PMDS], *D1* [PMDS], *D2* [PMDS], *E1* [FP], *E2* [FP]. **p114: Marsh Bluetail** habitat [PMDS]. **p115: Marsh Bluetail** *A1* [FP], *A2* [FP], *B1* [FP], *B2* [FP], *C1* [FP], *C/D2* [DSm], *D1* [FP]. **p116: Oasis Bluetail** habitat [GR/CC]. **p117: Oasis Bluetail** *A1* [FP], *A2* [FP], *B1* [FP], *B2* [FP], *C1* [FP], *C/D2* [FP], *D1* [FP]. **p118: Citrine Forktail** habitat [PV/CC]. **p119: Citrine Forktail** *A1* [FP], *A2* [FP], *B1* [FP], *B2* [FP], *C1* [FP], *C2* [ERN], *D1* [FP]. **p120: Sedgling** habitat [CB]. **p121: Sedgling** *A1* [CB], *A2* [FP], *B1* [FP], *B2* [FP], *C1* [FP], *C2* [ERN], *D1* [FP], *D1* [FP].

INTRODUCTION TO DRAGONFLIES (ANISOPTERA) **p122: Migrant Hawker** [PR]; **Western Spectre** [ERN]; **Blue Emperor** [A&GS]. **p123: Common Clubtail** [JT]; **Small Pincertail** [WL/Agami]; **Common Goldenring** [FP]; **Ringed Cascader** [PMDS]; **Splendid Cruiser** [FP]. **p124: Downy Emerald** [PR]; **Eurasian Baskettail** [ERN]; **Broad-bodied Chaser** [A&GS]; **Keeled Skimmer** [A&GS]; **Small Whiteface** [CFi/CC]. **p125: Common Darter** [DSm]; **Violet Dropwing** [DSm]; **Black Percher** [PW]; **Northern Banded Groundling** [FP]; **Wandering Glider** [A&GS].

HAWKERS, SPECTRES & EMPERORS **p126: Blue Hawker** [PR]. **p127: Migrant Hawker** [FS/Agami]; **Western Spectre** [ERN]; **Blue Emperor** [FP]. **p128: Hairy Hawker** [ERN]; **Migrant Hawker** [TS]; **Blue-eyed Hawker** [FS/Agami]; **Blue Hawker** [FP]; **Green Hawker** [CB]; **Azure Hawker** [FP]; **Moorland Hawker** [DSm]; **Bog Hawker** [ERN]; **Siberian Hawker** [ERN]; **Baltic Hawker** [FP]; **Brown Hawker** [ERN]; **Green-eyed Hawker** [FP]. **p129: Eastern Spectre** [FP]; **Western Spectre** [ERN]; **Cretan Spectre** [FP]; **Blue Emperor** [FP]; **Common Green Darner** [JJ]; **Lesser Emperor** [CB]; **Vagrant Emperor** [FP]; **Magnificent Emperor** [FP]. **p130: Hairy Hawker** [DSm]; **Migrant Hawker** [FS/Agami]; **Blue-eyed Hawker** [RG]; **Blue Hawker** [FP];

Green Hawker [ERN]; **Azure Hawker** [FP]; **Moorland Hawker** [WL/Agami]; **Bog Hawker** [ERN]; **Siberian Hawker** [FP]; **Baltic Hawker** [HS]; **Brown Hawker** [ERN]; **Green-eyed Hawker** [WL/Agami]. **p131: Eastern Spectre** [ERN]; **Western Spectre** (short appendages) [FP]; **Western Spectre** (long appendages) [FP]; **Cretan Spectre** [CB]; **Blue Emperor** [WL/Agami]; **Common Green Darner** [DP]; **Lesser Emperor** [FP]; **Vagrant Emperor** [FP]; **Magnificent Emperor** [BL]. **p132: Hairy Hawker** [WL/Agami]; **Migrant Hawker** [PR]; **Blue-eyed Hawker** [TS]; **Blue Hawker** [ERN]; **Green Hawker** [ERN]; **Azure Hawker** [AM]. **p133: Moorland Hawker** [PR]; **Bog Hawker** [FP]; **Siberian Hawker** [ERN]; **Baltic Hawker** [ERN]; **Brown Hawker** [PR]; **Green-eyed Hawker** [FP]. **p134: Eastern Spectre** [DSm]; **Western Spectre** [ERN]; **Cretan Spectre** [CB]. **p135: Blue Emperor** [A&GS]; **Common Green Darner** [MK]; **Lesser Emperor** [FP]; **Vagrant Emperor** [TS]; **Magnificent Emperor** [PC]. **p136: Hairy Hawker** habitat [DSm], inset (flight) [WL/Agami]. **p137: Hairy Hawker** *A1* [WL/Agami], *A2* [ERN], *B1* [WL/Agami], *B2* [DSm], *C1* [CB], *C2* [DSm]. **p138: Migrant Hawker** habitat [CB], inset (flight) [PR]. **p139: Migrant Hawker** *A1* [FP], *A2* [FS/Agami], *B2* [TS], *C1* [A&GS], *C2* [FS/Agami]. **p140: Blue-eyed Hawker** habitat [DSm], inset (flight) [TS]. **p141: Blue-eyed Hawker** *A1* [RG], *B1* [DKP], *C1* [PC], *D1* [FP], *D2* [DSm], *D3* [DSm], *D4* [ERN]. **p142: Blue Hawker** habitat [CB], inset (flight- top) [ERN], inset (flight -bottom) [DSm]. **p143: Blue Hawker** *A1* [ERN], *A2* [FP], *B2* [ERN], *C1* [A&GS], *C2* [DSm]. **p144: Green Hawker** habitat [CB], inset (flight) [ERN]. **p145: Green Hawker** *A1* [CB], *A2* [CB], *B1* [CB], *B2* [ERN], *C1* [CB]. **p146: Azure Hawker** habitat [CB], inset (flight) [AM]. **p147: Azure Hawker** *A1* [ERN], *A2* [FP], *B1* [FP], *B2* [FP], *C1* [ERN]. **p148: Moorland Hawker** habitat [CB], inset (flight) [PR]. **p149: Moorland Hawker** *A1* [A&GS], *A2* [DSm], *B2* [DSm], *C1* [A&GS], *C2* [FS/Agami], inset ('face') [PR], inset (top of frons) [DSm], inset (back of head) [DSm]. **p150: Bog Hawker** habitat [ERN], inset (flight-top) [FP], inset (flight-bottom) [PC]. **p151: Bog Hawker** *A1* [WL/Agami], *A2* [ERN], *B1* [ERN], *B2* [ERN], inset ('face') [FP], inset (top of frons) [PC]. **p152: Siberian Hawker** habitat [CB], inset (flight) [ERN]. **p153: Siberian Hawker** *A1* [ERN], *A2* [ERN], *B1* [FP], *B2* [FP], inset (thorax) [ERN]. **p154: Baltic Hawker** habitat [DSm], inset (flight) [ERN]. **p155: Baltic Hawker** *A1* [FP], *A2* [FP], *B1* [CB], *C1* [MF], *C2* [HS]. **p156: Brown Hawker** habitat [DSm], inset (flight) [PR]. **p157: Brown Hawker** *A1* [ERN], *A2* [PR], *B1* [WL/Agami], *B2* [ERN]. **p158: Green-eyed Hawker** habitat [DSm], inset (flight) [FP]. **p159: Green-eyed Hawker** *A1* [FP], *A2* [FP], *B1* [DSm], *B2* [WL/Agami], *C1* [WL/Agami], *C2* [RW/Alamy]. **p160: Eastern Spectre** habitat [CB], inset (flight) [DSm]. **p161: Eastern Spectre** *A1* [DSm], *A2* [FP], *B1* [PC], *B2* [PC], *C1* [DSm], *C2* [ERN]. **p162: Western Spectre** habitat [CB], inset (flight) [ERN]. **p163: Western Spectre** *A1* [ERN], *A2* [ERN], *B2* [FP], *C1* [FP], *C2* [FP]. **p164: Cretan Spectre** habitat [CB], inset (flight) [CB]. **p165: Cretan Spectre** *A1* [CB], *A2* [FP], *B1* [CB], *B2* [CB], inset (head) [CB]; **Western Spectre** inset (head) [ERN]. **p166: Blue Emperor** habitat [CB], inset (flight) [A&GS]. **p167: Blue Emperor** *A1* [FP], *A2* [FP], *B1* [DSm], *C1* [CB], *C2* [WL/Agami]. **p168: Common Green Darner** habitat [JC/CC], inset (flight) [MK]. **p169: Common Green Darner** *A1* [DP], *A2* [JJ], *B1* [DP], *B2* [DP], inset (top of frons) [JJ], inset (appendages) [JJ], inset (rear or head) [DP]; **Blue Emperor** inset (top of frons) [FP], inset (appendages) [FP], inset (rear or head) [WL/Agami]. **p170: Lesser Emperor** habitat [DSm], inset (flight) [FP]. **p171: Lesser Emperor** *A1* [CB], *A2* [CB], *B1* [PC], *B2* [JT], *C1* [TK]. **p172: Vagrant Emperor** habitat [CB], inset (flight) [TS]. **p173: Vagrant Emperor** *A1* [FP], *A2* [FP], *B1* [MB], *C1* [FP], *C2* [MB]. **p174: Magnificent Emperor** habitat [ERN], inset (flight) [PC]. **p175: Magnificent Emperor** *A1* [FP], *A2* [FP], *B1* [DSm], *B2* [BL].

CLUBTAILS, PINCERTAILS, SNAKETAIL, HOOKTAIL & BLADETAIL **p176: Clubtail** [WL/Agami]. **p177: Pincertail** [DSm]; **Snaketail** [FP]; **Hooktail** [PMDS]; **Bladetail** [WL/Agami]. **MALE APPENDAGES: Small Pincertail** [WL/Agami]; **Large Pincertail** [ERN]; **Faded Pincertail** [ERN]; **Waved Pincertail** [PC]; **Dark Pincertail** [PC]. **p178: Common Clubtail** habitat [A&GS]. **p179: Common Clubtail** *A1* [JT], *B1* [RG], *C1* [RG], *C2* [JT]. **p180: Turkish Clubtail** habitat [CB]. **p181: Turkish Clubtail** *A1* [FP], *B1* [DSm], *C1* [FP], *C2* [FP]. **p182: Yellow Clubtail** habitat [DSm]. **p183: Yellow Clubtail** *A1* [DSm], *B1* [DSm], *C1* [PMDS], *C2* [FP]. **p184: Pronged Clubtail** habitat [TDF/CC]. **p185: Pronged Clubtail** *A1* [FP], *B1* [FP], *C1* [FP], *C2* [PMDS]. **p186: Western Clubtail** habitat [CP]. **p187: Western Clubtail** *A1* [FP], *B1* [FP], *C1* [FP], *C2* [FP]. **p188: River Clubtail** habitat [CB]. **p189: River Clubtail** *A1* [FP], *B1* [CFi/CC], *C1* [FP], *C2* [TK]. **p190: Syrian Clubtail** habitat [DSm]. **p191: Syrian Clubtail** *A1* [FP], *B1* [PC], *C1* [FP], *C2* [PC]. **p192: Green Snaketail** habitat [DSm]. **p193: Green Snaketail** *A1* [FP], *A2* [FP], *B1* [FP], *B2* [DSm]. **p194: Small Pincertail** habitat [DSm]. **p195: Small Pincertail** *A1* [FP], *A2* [JT], *B1* [WL/Agami], *B2* [FP], *C1* [ERN], *C2* [FP], inset (appendages) [WL/Agami]. **p196: Large Pincertail** habitat [A&GS]. **p197: Large Pincertail** *A1* [A&GS], *B1* [FP], *C1* [FP], *C2* [ERN], inset (appendages) [ERN]. **p198: Faded Pincertail** habitat [DSm]. **p199: Faded Pincertail** *A1* [FP], *A2* [DSm], *B1* [FP], *B2* [FP], *C1* [FJCV], *C2* [ERN], inset (appendages) [ERN]. **p200: Waved Pincertail** habitat [DSm]. **p201: Waved Pincertail** *A1* [PC], *A2* [FP], *B1* [FP], *B2* [DSm], *C1* [FP], *C2* [FP]. **p202: Dark Pincertail** habitat [DSm]. **p203: Dark Pincertail** *A1* [PC], *A2* [FS/Agami], *B1* [PC], *B2* [FP], *C1* [PC]. **p204: Green Hooktail** habitat [CB]. **p205: Green Hooktail** *A1* [CB], *A/B2* [CB], *B1* [PMDS], *C1* [PMDS], *C2* [FP]. **p206: Bladetail** habitat [DSm]. **p207: Bladetail** *A1* [WL/Agami], *A2* [FP], *B2* [DSm], *C1* [PC], *C2* [FP].

GOLDENRINGS, CRUISER & CASCADER p208: Sombre Goldenring [FP]; **Common Goldenring** inset (head) [DSm]. **p209: Common Goldenring** [A&GS]; **Splendid Cruiser** [FP]; **Ringed Cascader** [FP]. **p210: MALES: Common Goldenring** northern [FP], southern [FP]; **Italian Goldenring** [FP]; **Balkan Goldenring** [FP]; **Turkish Goldenring** [FP]; **Sombre Goldenring** [FP]; **Greek Goldenring** [FP]; **Blue-eyed Goldenring** [FP]. **FEMALES: Common Goldenring** northern [FP], southern [FP]; **Italian Goldenring** [FP]; **Balkan Goldenring** [FP]; **Turkish Goldenring** [AC]; **Sombre Goldenring** [JBou]; **Greek Goldenring** [FP]; **Blue-eyed Goldenring** [YK]. **p211: Common Goldenring** inset (S1) [DSm], inset (male appendages) [DSm]; **Sombre Goldenring** inset (S1) [DSm], inset (male appendages) [DSm]. **p212: Common Goldenring** habitat [CB]. **p213: Common Goldenring** *A1* [FV/CC], *A/B2* [FP], *B1* [FP], *C1* [A&GS], *C2* [A&GS]. **p214: Italian Goldenring** habitat [CB]. **p215: Italian Goldenring** *A1* [FP], *A2* [FP], *B1* [PMDS], *B2* [FP], inset (head) [PMDS]. **p216: Balkan Goldenring** habitat [CB]. **p217: Balkan Goldenring** *A1* [DSm], *A2* [FP], *B1* [FP], *B2* [FP], inset (head) [DSm]. **p218: Turkish Goldenring** habitat [CB]. **p219: Turkish Goldenring** *A1* [DSm], *A2* [FP], *B1* [FP], *B2* [FP], inset (head) [DSm]. **p220: Sombre Goldenring** habitat [CB]. **p221: Sombre Goldenring** *A1* [DSm], *A2* [JB], *B1* [FP], *B2* [FP]. **p222: Greek Goldenring** habitat [CB]. **p223: Greek Goldenring** *A1* [FP], *A2* [FP], *B1* [FP], *B2* [FP]. **p224: Blue-eyed Goldenring** habitat [DSm]. **p225: Blue-eyed Goldenring** *A1* [JK], *A2* [JK], *B1* [FP], *B2* [FP]. **p226: Splendid Cruiser** habitat [DSm]. **p227: Splendid Cruiser** *A1* [FP], *A2* [FP], *B1* [FP], *B2* [FP]. **p228: Ringed Cascader** habitat [DSm], inset (flight) [PMDS]. **p229: Ringed Cascader** *A1* [PMDS], *A2* [PMDS], *B1* [PMDS], *B2* [PMDS].

EMERALDS p230: Downy Emerald [PR]; **Bulgarian Emerald** [FP]; **Brilliant Emerald** [PR]; **Balkan Emerald** [DSm]; **Yellow-spotted Emerald** [FP]; **Northern Emerald** [IC]; **Alpine Emerald** [TC]; **Treeline Emerald** [SK]; **Orange-spotted Emerald** [PMDS]. **p231: 'FACE': Downy Emerald** [DSm]; **Bulgarian Emerald** [DSm]; **Brilliant Emerald** [DSm]; **Balkan Emerald** [DSm]; **Yellow-spotted Emerald** [DSm]; **Northern Emerald** [SM]; **Alpine Emerald** [GSM/CC]; **Treeline Emerald** [MA]; **Orange-spotted Emerald** [DSm]. **MALE APPENDAGES: Downy Emerald** above [FP], side [DSm]; **Bulgarian Emerald** above [DSm], side [DSm]; **Brilliant Emerald** above [FP], side [DSm]; **Balkan Emerald** above [FP], side [FP]; **Yellow-spotted Emerald** above [FP], side [DSm]; **Northern Emerald** above [FP], side [FP]; **Alpine Emerald** above [CB], side [CB]; **Treeline Emerald** above [FP], side [FP]; **Orange-spotted Emerald** above [DSm], side [DSm]. **FEMALE VULVAR SCALE: Downy Emerald** [FP]; **Bulgarian Emerald** [FP]; **Brilliant Emerald** [FP]; **Balkan Emerald** [FP]; **Yellow-spotted Emerald** [DSm]; **Northern Emerald** [ERN]; **Alpine Emerald** [FP]; **Treeline Emerald** [FP]; **Orange-spotted Emerald** [FP]. **OTHER KEY FEATURES: Downy Emerald** S2/3 [JvD]; **Bulgarian Emerald** S2/3 [FP]; **Brilliant Emerald** thorax [TS]; **Balkan Emerald** thorax [FP]; **Northern Emerald** S3 [FP]; **Alpine Emerald** wing base [FP]; **Treeline Emerald** wing base [FP]. **p232: Downy Emerald** habitat [DSm], inset ('face') [DSm]. **p233: Downy Emerald** *A1* [FP], *A2* [JvD], *B1* [FP], *B2* [FP]. **p234: Bulgarian Emerald** habitat [DSm], inset ('face') [DSm]. **p235: Bulgarian Emerald** *A1* [FP], *B1* [FP], *B2* [FP], *C1* [FP]. **p236: Brilliant Emerald** habitat [CB], inset ('face') [DSm]. **p237: Brilliant Emerald** *A1* [FP], *A2* [FP], *B1* [DSm], *B2* [FP]. **p238: Balkan Emerald** habitat [CB], inset ('face') [DSm]. **p239: Balkan Emerald** *A1* [FP], *B1* [FP], *B2* [FP], *C1* [FP]. **p240: Yellow-spotted Emerald** habitat [DSm], inset ('face') [DSm]. **p241: Yellow-spotted Emerald** *A1* [CB], *A2* [FP], *B1* [FP], *C1* [FP], *C2* [FP], inset (vulvar scale) [DSm]. **p242: Northern Emerald** habitat [CB], inset ('face') [SM]. **p243: Northern Emerald** *A1* [FP], *B1* [FP], *B2* [FP], *C1* [FP]. **p244: Alpine Emerald** habitat [CB], inset ('face') [MA]. **p245: Alpine Emerald** *A1* [FP], *A2* [FP], *B1* [CB], *B2* [FP], inset (vulvar scale) [FP]. **p246: Treeline Emerald** habitat [CB], inset ('face') [SM]. **p247: Treeline Emerald** *A1* [FP], *B1* [FP], *B2* [FP], *C1* [FP]. **p248: Orange-spotted Emerald** habitat [DSm], inset ('face') [DSm]. **p249: Orange-spotted Emerald** *A1* [FP], *A2* [GG], *B1* [CB], *B2* [PMDS].

BASKETTAIL & CHASERS p250: Eurasian Baskettail habitat [DSm]. **p251: Eurasian Baskettail** *A1* [CB], *A2* [FP], *B2* [ERN], *C1* [FP], *C2* [FP]. **p252: Four-spotted Chaser** habitat [CB]. **p253: Four-spotted Chaser** *A1* [FP], *B1* [RG], *C1* [RG], *D1* [JT]. **p254: Broad-bodied Chaser** habitat [DSm]. **p255: Broad-bodied Chaser** *A1* [FP], *B1* [FP], *C1* [FP], *D1* [A&GS]. **p256: Blue Chaser** habitat [DSm]. **p257: Blue Chaser** *A1* [FP], *B1* [DSm], *C1* [DSm], *D1* [TK], *D2* [RG].

SKIMMERS p258: Black-tailed Skimmer [A&GS]; **Keeled Skimmer** [DSm]; **Skimmer** inset (wing) [FP]; **Darter** inset (wing) [PMDS]; **Dropwing** inset (wing) [ERN]. **p260: Black-tailed Skimmer** [A&GS]; **White-tailed Skimmer** [PC]; **Keeled Skimmer** [A&GS]; **Southern Skimmer** [A&GS]; **Yellow-veined Skimmer** [FP]; **Epaulet Skimmer** [DSm]; **Long Skimmer** [FP]; **Small Skimmer** [FP]; **Slender Skimmer** [PC]; **Blue Chaser** [A&GS]; **Lilypad Whiteface** [FS/Agami]. **p261: Black-tailed Skimmer** [DSm]; **White-tailed Skimmer** [A&GS]; **Keeled Skimmer** [A&GS]; **Southern Skimmer** [DSm]; **Yellow-veined Skimmer** [FP]; **Epaulet Skimmer** [FP]; **Long Skimmer** [FP]; **Small Skimmer** [FP]; **Slender Skimmer** [PC]; **Common Darter** [AI]; **Keeled Skimmer** pruinose [DSm]. **p262: Black-tailed Skimmer** habitat [DSm]. **p263: Black-tailed Skimmer** *A1* [MB], *A2* [A&GS], *B1* [JT], *B2* [JT], inset ('face') [A&GS]. **p264: White-tailed Skimmer** habitat [CB]. **p265: White-tailed Skimmer** *A1* [PC], *B1* [FP], *B2* [PC], *C1* [DSm], inset ('face') [DSm]. **p266: Keeled Skimmer** habitat [DSm]. **p267: Keeled Skimmer** *A1* [FP], *B1* [A&GS], *B2* [FP], *C1* [FP], *C2* [FP], inset (head/'face') [DSm]. **p268: Southern Skimmer** habitat [DSm]. **p269: Southern Skimmer** *A1* [JT], *B1* [ERN], *B2* [FP], *C1* [RG], inset (head/'face') [DSm]. **p270: Yellow-veined Skimmer** habitat [CB]. **p271: Yellow-veined Skimmer** *A1* [FP], *B1* [FP], *B2* [FP], *C1* [FP]. **p272: Desert Skimmer** habitat [JPB]. **p273: Desert Skimmer** *A1* [JPB], *B1* [JPB], inset (wing) [JPB]. **p274: Epaulet Skimmer** habitat [DSm]. **p275: Epaulet Skimmer** *A1* [FP], *A2* [DSm], *B1* [DSm], *C1* [FP], *C2* [DSm]. **p276: Long Skimmer** habitat [CB]. **p277: Long Skimmer** *A1* [FP], *A2* [FP], *B1* [FP], *B2* [FP]. **p278: Small Skimmer** habitat [DSm]. **p279: Small Skimmer** *A1* [FP], *A2* [WL/Agami], *B1* [FP], *B2* [FP], *C1* [FP], *C2* [FP]. **p280: Slender Skimmer** habitat [DSm]. **p281: Slender Skimmer** *A1* [ANSK], *B1* [PC], *B2* [PC], *C1* [FP].

WHITEFACES **p282: Yellow-spotted Whiteface** [CFi/CC]; **Small Whiteface** [CFi/CC]; **Lilypad Whiteface** [FS/Agami]. **p283: Small Whiteface** M [FP], F [ERN]; **Ruby Whiteface** M [JvD], F [JT]; **Yellow-spotted Whiteface** M [FP], F [FP]; **Dark Whiteface** M [TR], F [ERN]; **Lilypad Whiteface** M [FP], F [FS/Agami]; **Black Darter** M [CJS], F [BG]. **p284: Small Whiteface** habitat [DSm]. **p285: Small Whiteface** *A1* [FP], *A2* [RG], *B/C1* [ERN], *B2* [RG], *C2* [ERN]. **p286: Ruby Whiteface** habitat [CB]. **p287: Ruby Whiteface** *A1* [FP], *A2* [ERN], *B/C1* [ERN], *B2* [ERN], *C2* [JT]. **p288: Yellow-spotted Whiteface** habitat [DSm]. **p289: Yellow-spotted Whiteface** *A1* [ERN], *A2* [FP], *B2* [FP], *C1* [FP], *C2* [FP]. **p290: Dark Whiteface** habitat [ERN]. **p291: Dark Whiteface** *A1* [FP], *A2* [TR], *B1* [FP], *B2* [ERN], *C1* [FP], *C2* [BG], inset ('face') [FP]. **p292: Lilypad Whiteface** habitat [DSm]. **p293: Lilypad Whiteface** *A1* [FS/Agami], *A2* [FP], *B1* [FS/Agami], *B2* [WL/Agami], *C1* [PC], *C2* [FS/Agami].

DARTERS, SCARLET, DROPWINGS, PENNANT, PERCHER & GROUNDLING **p294: Skimmer** (wing) [FP]; **Darter** (wing) [PMDS]; **Dropwing** (wing) [ERN]; **Darter** [FP]. **p295: Broad Scarlet** [DSm]; **Violet Dropwing** [PC]; **Black Pennant** [DSm]; **Black Percher** [PW]; **Northern Banded Groundling** [FP]. **p296: Common Darter** [DSm]; **Orange-winged Dropwing** [DSm]; **Red-veined Dropwing** [PC]; **Violet Dropwing** [ERN]; **Indigo Dropwing** [DSm]; **Black Pennant** [DSm]; **Black Percher** [PW]; **Northern Banded Groundling** [PW]. **p297: Black Darter** [DSm]; **Broad Scarlet** [DSm]; **Ruddy Darter** [ERN]; **Common Darter** [A&GS]; **Red-veined Darter** [DSm]; **Moustached Darter** [FP]; **Yellow-winged Darter** [DSm]; **Spotted Darter** [PC]; **Banded Darter** [FP]; **Southern Darter** [PC]; **Island Darter** [PMDS]; **Desert Darter** [FP]. **p298: Black Darter** [ERN]; **Orange-winged Dropwing** [PMDS]; **Red-veined Dropwing** [FP]; **Violet Dropwing** [PC]; **Indigo Dropwing** [DSm]; **Black Pennant** [FP]; **Black Percher** [FP]; **Northern Banded Groundling** [FP]. **p299: Black Darter** [ERN]; **Broad Scarlet** [ERN]; **Ruddy Darter** [GH/CC]; **Common Darter** [FP]; **Red-veined Darter** [DSm]; **Moustached Darter** [ERN]; **Yellow-winged Darter** [FP]; **Spotted Darter** [DSm]; **Banded Darter** [DSm]; **Southern Darter** [ERN]; **Island Darter** [AETV]; **Desert Darter** [VCL]. **p300: Black Darter** habitat [CB], inset ('face') [PR]. **p301: Black Darter** *A1* [TK], *A2* [FP], *B1* [FP], *B/C2* [BG], *C1* [JT], *D1* [PR], *D2* [FP]. **p302: Common Darter** habitat [SS], inset ('face') [SN]. **p303: Common Darter** *A1* [A&GS], *A2* [RM], *B1* [J&CC], *B2* [FP], *C1* [FP], *C2* [PH]. **p304: Moustached Darter** habitat [CB], inset ('face') [DSm]. **p305: Moustached Darter** *A1* [FP], *A2* [FP], *B1* [FP], *B/C2* [FP], *C1* [DSm], *D1* [FP], *D2* [JT]. **p306: Island Darter** habitat [JBow], inset ('face') [PMDS]. **p307: Island Darter** *A1* [PMDS], *A2* [PMDS], *B1* [PMDS], *B2* [PMDS], *C1* [AETV], *C2* [J&CC]. **p308: Southern Darter** habitat [CB], inset ('face') [PC]. **p309: Southern Darter** *A1* [PC], *A2* [FP], *B1* [PC], *B2* [FS/Agami], *C1* [FP], *C2* [PC]. **p310: Desert Darter** habitat [CB], inset ('face') [FP]. **p311: Desert Darter** *A1* [FP], *A2* [FP], *B1* [VCL], *B2* [FP], *C1* [FP], *C2* [SO]. **p312: Ruddy Darter** habitat [DSm], inset ('face') [FP]. **p313: Ruddy Darter** *A1* [FP], *A2* [ERN], *B1* [FP], *B2* [A&GS], *C1* [DSm], *C2* [PC]. **p314: Spotted Darter** habitat [CB], inset ('face') [PC], inset (wing) [DSm]; **Ruddy Darter** inset (wing) [DSm]. **p315: Spotted Darter** *A1* [PC], *A2* [FP], *B1* [FS/Agami], *B2* [DSm], *C1* [PC], *C2* [DSm]. **p316: Red-veined Darter** habitat [DSm], inset ('face') [PC]. **p317: Red-veined Darter** *A1* [PC], *A2* [FP], *B1* [FS/Agami], *B/C2* [FP], *C1* [CFe/CC], *D1* [PC], *D2* [FP]. **p318: Yellow-winged Darter** habitat [CB], inset ('face') [FP]. **p319: Yellow-winged Darter** *A1* [FP], *A2* [FP], *B1* [FP], *B2* [FP], *C1* [FP], *C2* [RG]. **p320: Banded Darter** habitat [CB], inset ('face') [ERN]. **p321: Banded Darter** *A1* [FP], *A2* [FS/Agami], *B1* [FP], *B2* [FP], *C1* [ERN], *C2* [DSm]. **p322: Broad Scarlet** habitat [CB]. **p323: Broad Scarlet** *A1* [FP], *A2* [FP], *B1* [ERN], *B2* [ERN], *C1* [ERN], *C2* [ERN]. **p324: Orange-winged Dropwing** habitat [DSm]. **p325: Orange-winged Dropwing** *A1* [PMDS], *A2* [PMDS], *B2* [FP], *C1* [FP], *C2* [FP]. **p326: Red-veined Dropwing** habitat [DSm]. **p327: Red-veined Dropwing** *A1* [FP], *A2* [FP], *B1* [WL/Agami], *B2* [PMDS], *C1* [PC]. **p328: Violet Dropwing** habitat [CB]. **p329: Violet Dropwing** *A1* [ERN], *A2* [FP], *B1* [FP], *B2* [DSm], *C1* [ERN], *C2* [ERN]. **p330: Indigo Dropwing** habitat [CB]. **p331: Indigo Dropwing** *A1* [ERN], *A2* [DSm], *B1* [PC], *B2* [FP], *C1* [PC]. **p332: Black Pennant** habitat [CB]. **p333: Black Pennant** *A1* [DSm], *A2* [DSm], *B1* [FP], *B2* [FP], *C1* [PC], *C2* [DSm]. **p334: Black Percher** habitat [CB]. **p335: Black Percher** *A1* [FP], *A2* [FP], *B1* [FP], *B2* [FP], *C1* [DSm], *C2* [PW], *D1* [DSm], *D2* [FP]. **p336: Northern Banded Groundling** habitat [CB]. **p337: Northern Banded Groundling** *A1* [FP], *A2* [WL/Agami], *B1* [FP], *B2* [CB], *C1* [ERN].

GLIDERS **p338: Keyhole Glider** flight [MV]. **p339: Keyhole Glider** *A1* [NM], *B1* [PMDS], *C1* [PMDS]. **p340: Wandering Glider** habitat [DSp]. **p341: Wandering Glider** *A1* [A&GS], *A2* [PC], *B1* [FP], *C1* [FP].

p348: Blue-eyed Hawker [FP].

Index

This index includes the English and *scientific (in italics)* names of all the damselflies and dragonflies in this book, as well as the families and 'types' of Odonata.

Names in **bold black text** are the English names used throughout the book; alternative English names that are used elsewhere in Europe are shown in regular text. **Bold blue text** signifies families and (in CAPITALS) 'types' of Odonata.

Numbers in **bold** relate to introductory sections and the main species accounts. Numbers in regular text indicate pages where comparative tables or charts appear. Numbers in blue for families and genera refer to the *Checklist of species*.

A

Aeshna **126, 127,** 345
— *affinis* 128, 130, 132, **140**
— *caerulea* 128, 130, 132, **146**
— *crenata* 128, 130, 133, **152**
— *cyanea* 128, 130, 132, **142**
— *grandis* 128, 130, 133, **156**
— *isoceles* 128, 130, 133, **158**
— *juncea* 128, 130, 133, **148**
— *mixta* 128, 130, 132, **138**
— *serrata* 128, 130, 133, **154**
— *subarctica* 128, 130, 133, **150**
— *viridis* 128, 130, 132, **144**
Aeshnidae **126,** 345
Anax **126, 127,** 345
— *ephippiger* 129, 131, 135, **172**
— *immaculifrons* 129, 131, 135, **174**
— *imperator* 129, 131, 135, **166,** 169
— *junius* 129, 131, 135, **168**
— *parthenope* 129, 131, 135, **170**

B

BASKETTAIL 250
Baskettail, Eurasian 250
BLADETAIL **176, 177**
Bladetail 177, **206**
Blue-eye 65, 67, 95, **96**
BLUET **64, 66**
Bluet, Arctic 65–67, **76**
—, **Azure** 65–67, **70**
—, **Common** 65–67, **68**
—, **Crescent** 65–67, **80**
—, **Cretan** 65–67, **92**
—, **Dainty** 65–67, **88,** 90
—, **Dark** 65–67, **82**
—, **Irish** **80**
—, **Mediterranean** 65–67, **88, 90**
—, **Mercury** 65–67, **84**
—, **Ornate** 65–67, **86**
—, **Siberian** 65–67, **78**
—, **Spearhead** 65–67, **74**
—, **Variable** 65–67, **72**

BLUETAIL **94**
Bluetail, Common 95, **106**
—, **Common** **114**
—, **Dumont's** **104**
—, **Iberian** 95, **108**
—, **Island** 95, **110**
—, **Marsh** 95, **114**
—, **Oasis** 95, **116**
—, **Persian** 95, **104**
—, **Sahara** 95, **112**
—, **Small** 95, **102**
—, **Tropical** **114**
Bluetip, Common **106**
—, Small **102**
Boyeria **126, 127,** 345
— *cretensis* 129, 131, 134, **164**
— *irene* 129, 131, 134, **162,** 165
Brachythemis **294, 295,** 346
— *impartita* 296, 298, **336**
Brachytron **126, 127,** 345
— *pratense* 128, 130, 132, **136**
BRIGHTEYE **94**

C

Caliaeschna **126, 127,** 345
— *microstigma* 129, 131, 134, **160**
Calopterygidae **38,** 343
Calopteryx **38,** 343
— *haemorrhoidalis* 39, **46**
— *splendens* 39, **42**
— *virgo* 39, **40**
— *xanthostoma* 39, **44**
CASCADER **208, 209**
Cascader, Ringed **228**
Ceriagrion **50,** 344
— *georgifreyi* **56**
— *tenellum* 54, **56**
Chalcolestes **18,** 343
— *parvidens* 19, **32**
— *viridis* 19, **30**

CHASER ... 250
Chaser, Blue ... 256
—, Broad-bodied ... 254
—, Four-spotted ... 252
—, Scarce ... 256
CLUBTAIL ... 176
Clubtail, Common ... 176, 178
—, Green ... 192
—, Pronged ... 176, 184
—, River ... 176, 188
—, Syrian ... 176, 190
—, Turkish ... 176, 180
—, Western ... 176, 186
—, Yellow ... 176, 182
—, Yellow-legged ... 188
Coenagrion ... **64, 66**, 344
— *armatum* ... 65–67, **82**
— *caerulescens* ... 65–67, 88, **90**
— *hastulatum* ... 65–67, **74**
— *hylas* ... 65–67, **78**
— *intermedium* ... 65–67, **92**
— *johanssoni* ... 65–67, **76**
— *lunulatum* ... 65–67, **80**
— *mercuriale* ... 65–67, **84**
— *ornatum* ... 65–67, **86**
— *puella* ... 65–67, **70**
— *pulchellum* ... 65–67, **72**
— *scitulum* ... 65–67, **88**, 90
Coenagrionidae ... **50**, 64, 66, **94**, 344
Cordulegaster ... **208, 209**, 346
— *bidentata* ... 210–211, **220**
— *boltonii* ... 210–211, **212**
— *helladica* ... 210–211, **222**
— *heros* ... 210–211, **216**
— *insignis* ... 210–211, **224**
— *picta* ... 210–211, **218**
— *trinacriae* ... 210–211, **214**
Cordulegastridae ... **208**, 346
Cordulia ... **230**, 346
— *aenea* ... 230–231, **232**
Corduliidae ... **230, 250**, 346
Corduliochlora ... **230**, 346
— *borisi* ... 230–231, **234**
Crocothemis ... **294, 295**, 346
— *erythraea* ... 297, 299, **322**
CRUISER ... **208, 209**
Cruiser, Splendid ... 226

D
Damsel, Common Winter ... 19, 34
Damsel, Greek Red ... 52
—, Large Red ... 50, 53
DAMSEL, RED ... 50
Damsel, Siberian Winter ... 19, 36

Damsel, Small Red ... 54, 56
—, Turkish Red ... 56
DAMSEL, WINTER ... 18, 19
Damselfly, Azure ... 70
—, Bilek's ... 78
DAMSELFLY, 'BLUE-TAILED' ... 94
Damselfly, Blue-tailed ... 106
—, Common Blue ... 68
—, Dainty ... 88
—, Dark Emerald ... 28
—, Emerald ... 20
—, Frey's ... 78
—, Goblet-marked ... 96
—, Irish ... 80
—, Large Red ... 50
—, Norfolk ... 82
—, Northern ... 74
—, Orange White-legged ... 58
—, Pygmy ... 120
—, Red-eyed ... 98
—, Scarce Blue-tailed ... 102
—, Scarce Emerald ... 22
—, Small Emerald ... 24
—, Small Red ... 54
—, Small Red-eyed ... 100
—, Southern ... 84
—, Southern Emerald ... 26
—, Variable ... 72
—, White-legged ... 62
—, Willow Emerald ... 30
—, Winter ... 34
Darner, Common Green ... 129, 131, 135, 168
DARTER ... 294
Darter, Banded ... 297, 299, 320
—, Black ... 297, 299, 300
—, Common ... 297, 299, 302
—, Dainty White-faced ... 292
—, Desert ... 297, 299, 310
—, Eastern White-faced ... 290
—, Highland ... 302
—, Island ... 297, 299, 306
—, Large White-faced ... 288
—, Marshland ... 314
—, Moustached ... 297, 299, 304
—, Northern White-faced ... 286
—, Red-veined ... 297, 299, 316
—, Ruddy ... 297, 299, 312, 314
—, Scarlet ... 322
—, Southern ... 297, 299, 308
—, Spotted ... 297, 299, 314
—, Vagrant ... 304
—, Violet-marked ... 328
—, White-faced ... 284
—, Yellow-winged ... 297, 299, 318

DEMOISELLE..38
Demoiselle, Banded.............................39, 42
—, Beautiful...39, 40
—, Copper..39, 46
—, Mediterranean......................................46
—, Western...39, 44
—, Yellow-tailed...44
Diplacodes..............................294, 295, 346
— *lefebvrii*.............................296, 298, 334
Dragonfly, Blue-eyed Hook-tailed...........196
—, Club-tailed..178
—, Emperor..166
—, Golden-ringed.....................................212
—, Green-eyed Hook-tailed......................194
—, Hairy..136
—, Two-toothed Golden-ringed................220
DROPWING.................................294, 295
Dropwing, Indigo................296, 298, 330
—, Kirby's...324
—, Orange-winged................296, 298, 324
—, Red-veined......................296, 298, 326
—, Violet..............................296, 298, 328

E
EMERALD...230
Emerald, Alpine...................230–231, 244
—, Balkan............................230–231, 238
—, Brilliant..........................230–231, 236
—, Bulgarian........................230–231, 234
—, Downy.............................230–231, 232
—, Moorland...242
—, Northern.........................230–231, 242
—, Orange-spotted...............230–231, 248
—, Treeline...........................230–231, 246
—, Yellow-spotted................230–231, 240
EMPEROR.................................126, 127
Emperor, Blue........129, 131, 135, 166, 169
—, Fiery...174
—, Lesser................129, 131, 135, 170
—, Magnificent........129, 131, 135, 174
—, Vagrant..............129, 131, 135, 172
—, Yellow-winged.....................................170
Enallagma..............................64, 66, 344
— *cyathigerum*....................65–67, 68
Epallage....................................48, 343
— *fatime*..48
Epitheca....................................250, 346
Epitheca bimaculata...............................250
Erythromma..............................94, 344
— *lindenii*.......................65, 67, 95, 96
— *najas*...95, 98
— *viridulum*....................................95, 100
Euphaeidae...................................48, 343

F
FEATHERLEG..58
Featherleg, Blue.................................60, 62
—, Orange..58
—, White...60, 62
Forktail, Citrine.................................95, 118

G
GLIDER...338
Glider, Keyhole..338
—, Wandering...340
—, Wheeling...338
GOLDENRING...............................208, 209
Goldenring, Balkan............210–211, 216
—, Blue-eyed.......................210–211, 224
—, Common.........................210–211, 212
—, Greek..............................210–211, 222
—, Italian............................210–211, 214
—, Sombre...........................210–211, 220
—, Turkish...........................210–211, 218
Gomphidae...............................176, 345
Gomphus.................................176, 345
— *graslinii*.........................176, 184
— *pulchellus*......................176, 186
— *schneiderii*.....................176, 180
— *simillimus*......................176, 182
— *vulgatissimus*.................176, 178
GROUNDLING.............................294, 295
Groundling, Northern Banded...296, 298, 336

H
Hawk, Green Marsh................................280
HAWKER.................................126, 127
Hawker, Amber-winged...........................156
—, Autumn...138
—, Azure...............128, 130, 132, 146
—, Baltic...............128, 130, 133, 154
—, Blue.................128, 130, 132, 142
—, Blue-eyed.........128, 130, 132, 140
—, Bog..................128, 130, 133, 150
—, Brown...............128, 130, 133, 156
—, Common..148
—, Dusk...162
—, Green...............128, 130, 132, 144
—, Green-eyed.......128, 130, 133, 158
—, Hairy...............128, 130, 132, 136
—, Migrant............128, 130, 132, 138
—, Moorland..........128, 130, 133, 148
—, Norfolk...158
—, Siberian...........128, 130, 133, 152
—, Southern..142
—, Southern Migrant................................140
—, Spring...136
—, Subarctic..150

HOOKTAIL...**176, 177**
Hooktail, Blue-eyed.............................**196**
—, **Green**....................................177, **204**
—, Green-eyed..................................**194**

I
Ischnura...**94**, 344
— *elegans*.....................................95, **106**
— *fountaineae*...............................95, **116**
— *genei*...95, **110**
— *graellsii*....................................95, **108**
— *hastata*......................................95, **118**
— *intermedia*.................................95, **104**
— *pumilio*......................................95, **102**
— *saharensis*.................................95, **112**
— *senegalensis*..............................95, **114**

J
Jewelwing, Banded................................**42**
—, Beautiful...**40**

L
Lestes...**18**, 343
— *barbarus*......................................19, **26**
— *dryas*...19, **22**
— *macrostigma*...............................19, **28**
— *sponsa*...19, **20**
— *virens*..19, **24**
Lestidae...**18**, 343
Leucorrhinia...............................**282**, 346
— *albifrons*...................................283, **290**
— *caudalis*....................................283, **292**
— *dubia*..283, **284**
— *pectoralis*.................................283, **288**
— *rubicunda*.................................283, **286**
Libellula.......................................**250**, 347
— *depressa*....................................**254**
— *fulva*...**256**
— *quadrimaculata*..........................**252**
Libellulidae
................**208, 250, 258, 282, 294, 338**, 346
Lindenia..................................**176, 177**, 345
— *tetraphylla*...............................177, **206**

M
Macromia.............................**208, 209**, 346
— *splendens*...................................**226**
Macromiidae.............................**208**, 346

N
Nehalennia...................................**94**, 344
— *speciosa*....................................95, **120**

O
ODALISQUE..**48**
Onychogomphus.....................**176, 177**, 345
— *assimilis*....................................177, **202**
— *costae*.......................................177, **198**
— **flexuosus**.................................177, **200**
— *forcipatus*.................................177, **194**
— *uncatus*.....................................177, **196**
Ophiogomphus.......................**176, 177**, 345
— *cecilia*.......................................177, **192**
Orthetrum.....................................**258**, 347
— *albistylum*.............................259–261, **264**
— *brunneum*..............................259–261, **268**
— *cancellatum*...........................259–261, **262**
— *chrysostigma*.........................259–261, **274**
— *coerulescens*..........................259–261, **266**
— *nitidinerve*............................259–261, **270**
— *ransonnetii*...............................259, **272**
— *sabina*...................................259–261, **280**
— *taeniolatum*...........................259–261, **278**
— *trinacria*................................259–261, **276**
Oxygastra.....................................**230**, 346
— *curtisii*..................................230–231, **248**

P
Pantala...**338**, 347
— *flavescens*....................................**340**
Paragomphus.........................**176, 177**, 346
— *genei*...177, **204**
PENNANT....................................**294, 295**
Pennant, Black...................296, 298, **332**
PERCHER......................................**294, 295**
Percher, Black....................296, 298, **334**
PINCERTAIL.................................**176, 177**
Pincertail, Dark................................177, **202**
—, **Faded**.....................................177, **198**
—, **Large**.....................................177, **196**
—, **Small**.....................................177, **194**
—, **Waved**...................................177, **200**
Platycnemididae..............................**58**, 343
Platycnemis....................................**58**, 343
— *acutipennis*....................................**58**
— *latipes*......................................**60**, 62
— *pennipes*...................................**60, 62**
Pyrrhosoma...................................**50**, 344
— *elisabethae*....................................**52**
— *nymphula*...................................**50**, 53

R
Redeye, Large...................................95, **98**
—, **Small**......................................95, **100**
Redtail, Small......................................**54**
—, Spring...**50**

S
SCARLET...........................**294, 295**
Scarlet, Broad..............297, 299, **322**
SEDGLING.................................**94**
Sedgling.........................95, **120**
Selysiothemis............**294, 295**, 347
— nigra.....................296, 298, **332**
SKIMMER...............................**258**
Skimmer, Black-tailed.......259–261, **262**
—, Desert..................259, **272**
—, Epaulet.............259–261, **274**
—, Globe................................**340**
—, Green...............................**280**
—, Heathland.........................**266**
—, Keeled............259–261, **266**
—, Long...............259–261, **276**
—, Ransonnet's........................**272**
—, Slender...........259–261, **280**
—, Small.............259–261, **278**
—, Sombre.............................**280**
—, Southern........259–261, **268**
—, White-tailed....259–261, **264**
—, Yellow-veined...259–261, **270**
SNAKETAIL....................**176, 177**
Snaketail, Green............177, **192**
Somatochlora..................**230**, 346
— alpestris..........230–231, **244**
— arctica............230–231, **242**
— flavomaculata...230–231, **240**
— meridionalis.....230–231, **238**
— metallica.........230–231, **236**
— sahlbergi........230–231, **246**
SPECTRE........................**126, 127**
Spectre, Cretan....129, 131, 134, **164**
—, Eastern.......129, 131, 134, **160**
—, Western....129, 131, 134, **162**, 165
Spiketail, Golden-ringed.............**212**
SPREADWING..........................**18**
Spreadwing, Common............19, **20**
—, Dark...........................19, **28**
—, Eastern Willow..............19, **32**
—, Migrant.......................19, **26**
—, Robust........................19, **22**
—, Small.........................19, **24**
—, Turlough.........................**22**
—, Western Willow..............19, **30**

Stylurus..........................**176**, 346
— flavipes.................176, **188**
— ubadschii...............176, **190**
Sympecma....................**18, 19**, 343
— fusca......................19, **34**
— paedisca...................19, **36**
Sympetrum.....................**294**, 347
— danae...............297, 299, **300**
— depressiusculum...297, 299, **314**
— flaveolum.........297, 299, **318**
— fonscolombii.....297, 299, **316**
— meridionale.....297, 299, **308**
— nigrescens.........................**302**
— nigrifemur........297, 299, **306**
— pedemontanum...297, 299, **320**
— sanguineum....297, 299, **312**, 314
— sinaiticum.......297, 299, **310**
— striolatum........297, 299, **302**
— vulgatum.........297, 299, **304**

T
Tramea..........................**338**, 347
— basilaris............................**338**
Trithemis....................**294, 295**, 347
— annulata..........296, 298, **328**
— arteriosa..........296, 298, **326**
— festiva............296, 298, **330**
— kirbyi.............296, 298, **324**

W
WHITEFACE..........................**282**
Whiteface, Dark..............283, **290**
—, Lilypad..................283, **292**
—, Ruby.....................283, **286**
—, Small....................283, **284**
—, Yellow-spotted...........283, **288**

Z
Zygonyx....................**208, 209**, 347
— torridus............................**228**